C000128684

X-PLANES 11

JET PROTOTYPES OF WORLD WAR II

Gloster, Heinkel, and Caproni Campini's wartime jet programmes

Tony Buttler

SERIES EDITOR TONY HOLMES

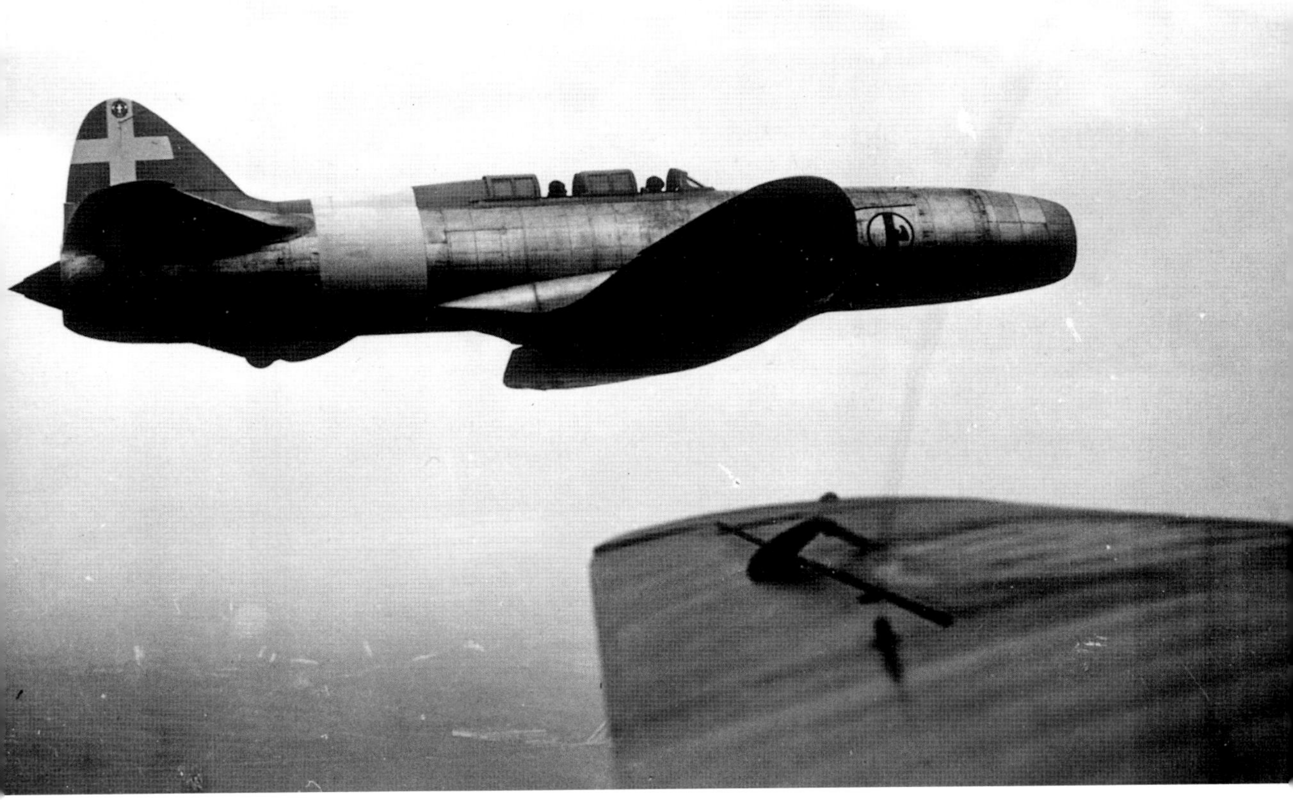

OSPREY
PUBLISHING

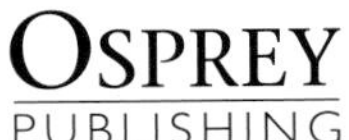

OSPREY PUBLISHING
Bloomsbury Publishing Plc
PO Box 883, Oxford, OX1 9PL, UK
1385 Broadway, 5th Floor, New York, NY 10018, USA
E-mail: info@ospreypublishing.com
www.ospreypublishing.com

OSPREY is a trademark of Osprey Publishing Ltd

First published in Great Britain in 2019

A catalogue record for this book is available from the British Library.

ISBN: PB 9781472835987; eBook 9781472835994;
ePDF 9781472835970; XML 9781472836007

A CIP catalogue record for this book is available from the British Library

19 20 21 22 23 10 9 8 7 6 5 4 3 2 1

Edited by Tony Holmes
Artwork by Adam Tooby
Index by Fionbar Lyons
Originated by PDQ Digital Media Solutions, UK
Printed in China through World Print Ltd

Osprey Publishing supports the Woodland Trust, the UK's leading woodland conservation charity.

Acknowledgements
I would like to thank the following for their vital contributions to this work, both with reference materials and illustrations – Dr.-Ing. Giorgio Apostolo, Phil Butler, Marco Comelli, Tim Kershaw and the Jet Age Museum archive, Henry Matthews, Wolfgang Muehlbauer and the staff of the British National Archives at Kew.

Front Cover
The third prototype Heinkel He 280 is captured flying at high speed through a beautiful cloudscape during one of its test flights. This artwork by Adam Tooby was specially commissioned for this book.

X PLANES

CONTENTS

The extraordinary Whittle Unit, which was successfully run on the bench for the first time on 12 April 1937. Although clearly not suitable for installation in an aeroplane, this engine provided Frank Whittle and his team with important knowledge and experience in jet operation prior to the construction of their first aero engine. (Author's Collection)

CHAPTER ONE

JET PROPULSION

During the six long years of World War II the major nations involved developed a string of outstanding piston-engined fighter aircraft. Production types like the late marks of the Focke-Wulf Fw 190, North American P-51 Mustang and Supermarine Spitfire, plus various advanced prototypes flown towards the end of the conflict, represented a peak of development for piston-engined combat aircraft and their powerplants. Yet behind the scenes, even before war had broken out, programmes for more advanced forms of aircraft propulsion were underway in Britain, Germany and Italy. These would result in some fascinating prototypes and, although the approach taken by the latter nation would come to nothing, the development of the jet engine by the remaining two countries would lay the ground for all of the high performance military and civil aircraft produced since 1945.

GREAT BRITAIN

The story of the development of the jet engine in Great Britain is well known, having been covered in many books and articles. It was also a very political affair, but that side has been ignored in the brief account given here.

The jet was the brainchild of Frank Whittle. In fact, quite rightly, he is credited with single-handedly inventing the turbojet engine. Whittle worked on his early theories of jet propulsion during the first half of the 1930s while serving as an RAF officer, and he patented his engine design in 1930. Frenchman Maxime Guillaume had actually filed a patent to use an axial-flow gas turbine to power an aircraft as early as

Sir Frank Whittle photographed in 1949 in front of the de Havilland Comet, the world's first commercial jet airliner. When Whittle initially formed his theories for jet propulsion he had civil aircraft in mind, not military types. He received his knighthood, and retired from the RAF, in 1948. (BAE Systems Heritage)

1921, but at that time the state of the art in compressors and materials made such a concept impossible to bring to fruition.

In 1936, and at this stage without any official (Ministry) backing, Whittle and two colleagues formed Power Jets Ltd at Rugby, in Warwickshire, in premises hired from the British Thomson-Houston Company (BTH). Despite a lack of finance, its first jet engine, the Whittle Unit (WU), was successfully run on the bench on 12 April 1937. However, the WU was rather large and in no way could be described as a 'flyable' engine, but its success did at last stimulate some official interest. As regards his RAF career, Whittle had now been put on the 'Special Duty List' because he was involved in research work.

In 1938 Power Jets moved to BTH's nearby Ladywood foundry in Lutterworth, Leicestershire, and in 1939, now with an Air Ministry contract secured, the firm began the development of the W1 – the first Power Jets aero engine. By 1940 the firm was employing about 25 people, so this landmark effort was undertaken by a very small team. It was the W1 that powered the Gloster E.28/39 (the subject of Chapter Five) on its maiden flight in 1941.

An important decision made by Whittle was to keep his engine as simple as possible, in particular through the use of the centrifugal compressor arrangement with a single large aluminium impeller to

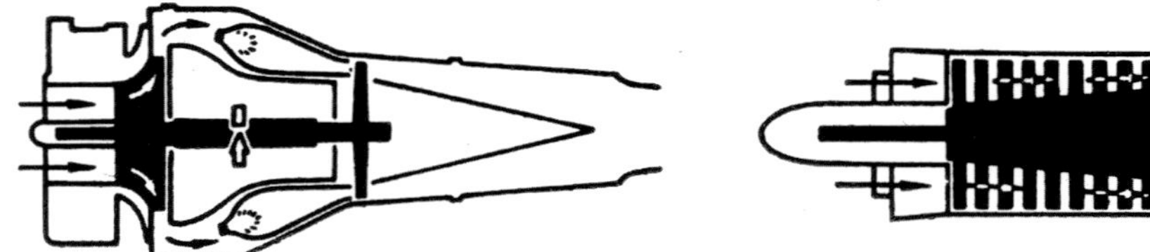

Two approaches to jet engine design. The first drawing shows an example of a centrifugal jet – in fact a de Havilland Ghost, a type which first ran in 1945. This had a single-sided impellor for its compressor, separate combustion chambers and a single-stage turbine. The second is an axial jet, in this particular case the small low-thrust Armstrong Siddeley Viper from the early 1950s. This featured a nine-stage compressor (nine discs, each with their own set of blades), an annular combustion chamber and a single-stage turbine. (Author's Collection)

compress the air before it passed through into the combustion chamber. The alternative was the axial arrangement in which the compressor had discs, with sets of small blades fitted around their circumference. This was the approach adopted by Germany and also pushed in Britain by specialists (specifically Dr A. A. Griffith) at the Royal Aircraft Establishment (RAE) at Farnborough, in Hampshire.

Another factor that helped Whittle was the availability of suitable metals capable of coping with the severe heat conditions and stresses found inside a jet engine. A company called High Duty Alloys based at Slough, in Berkshire, and Redditch, in Worcestershire, in collaboration with Rolls-Royce, produced special aluminium alloys under the trade name Hiduminium that possessed properties ideal for operation within a jet engine compressor. Furthermore, during 1940, Henry Wiggin of Hereford, in Herefordshire, developed the Nimonic family of nickel-based 'superalloys' that could retain their strength at the very high temperatures (to begin with around 800°C) produced within a jet's turbine.

The creation of these metals gave Whittle a big advantage over his German counterparts who had no access to large supplies of nickel. Instead, they had to use stainless steel alloys in their jet turbines (Tinidur for example) that could not cope with the temperatures to which they were exposed. This in turn meant that these engines had

A forged compressor impellor ready for final machining. When the engine was running this would rotate at high speed to compress the incoming air. This particular example was manufactured for the de Havilland H.1 Goblin engine. (Author's Collection)

A bladed turbine disc – an axial compressor disc would look quite similar. The spinning blades fitted to the edge of the disc had aerofoil sections to help compress the air. Each disc and its blades were known as a 'stage'. (Author's Collection)

to be overhauled after very short periods, at most after just 25 hours. Tinidur was also very difficult to work and it could not be welded.

The Rover Car Company was given the task of putting the Whittle engine into production, but its relationship with Power Jets was never harmonious and in 1943 Whittle's company began working instead with Rolls-Royce. The next Power Jets design was called the W2, and with Rolls-Royce's help in development this became the RB.23 Welland that powered the earliest marks of the Gloster Meteor, Britain's first jet fighter.

The existence of the jet engine was finally made public in newspaper reports published in January 1944. Two months later Power Jets was officially nationalised to become Power Jets (Research and Development) Ltd. Then in July 1946 the organization was merged with the gas turbine division of RAE Farnborough to form the National Gas Turbine Establishment (NGTE). Sadly, after these moves, Whittle turned his back on the industry. In 1948 he was knighted for his work.

It must be mentioned that by the late 1940s five major British companies had all become involved in the development and manufacture of jet engines. In 1941 de Havilland began a jet fighter project which became the DH.100 Vampire, Britain's second jet fighter. This was powered by de Havilland's own H.1 Goblin centrifugal engine designed by Frank Halford. Compared to Whittle's designs, the H.1 was 'cleaned up' in that it used a single-sided compressor with the inlet at the front and a 'straight through' layout with the combustion chambers exhausting directly onto the turbine. Whittle had used a double-sided compressor and a 'reverse flow' layout that piped the hot air back to the middle of the engine. This second feature in effect 'folded' the airflow and thus reduced the engine's length. The H.1 first ran in April 1942.

Bristol Engines and Armstrong Siddeley also began their own research programmes, while as early as 1939 Metropolitan-Vickers set up a project to develop an axial-flow engine that it called the F.2 Beryl. It is no exaggeration to say that Whittle's work caused a revolution within the British aero engine manufacturing industry. He died in 1996, which was late enough for him to witness the global growth of air travel – a change made possible almost entirely by his creation of the jet engine.

GERMANY

The development of the jet engine in Germany, once the programme was up and running, for a period ran near concurrently with that in Britain (indeed a fine account of the progress of both, placed alongside one another, can be found in Glyn Jones' book *The Jet Pioneers*). The German pioneer here was physicist Hans Joachim Pabst von Ohain, who duly became the designer of the first *operational* jet engine – that is, the first jet engine to take an aircraft into the air. Unfortunately, von Ohain's creations would never power a production aeroplane.

In 1935, while he was based at the University of Göttingen, von Ohain patented a turbojet unit that had a single-stage centrifugal compressor and a single-stage radial inflow turbine. At this point he was some years behind Whittle but, unlike the rather fragmented British gas turbine design efforts up until around the start of World War II, Germany's jet engine and jet aircraft projects would prove to be rather better funded and coordinated and the gap was closed very quickly. Indeed, by the beginning of the war, Germany had taken the lead.

In February 1936 Robert Pohl, the director of the Physical Institute at Göttingen and for whom von Ohain was working, wrote a letter to German aircraft designer Ernst Heinkel outlining von Ohain's research. After a meeting with Heinkel and his engineers on 17 March, von Ohain and his mechanic colleague Max Hahn began to work for Heinkel at his facility at Marienehe airfield, near Rostock in April. In late summer 1936 they started to build the relatively simple HeS 1 (Heinkel-Strahltriebwerk 1), primarily in sheet metal because this was to serve as a demonstration engine, not an aero engine. Designed to run on hydrogen gas, it was completed in March 1937 and first ran two weeks before Whittle began his first trials with the WU. Von Ohain had caught up, and in general the HeS 1's test runs were successful, with the unit producing some 551lb of thrust. Parts of its structure were damaged, however, by the extreme heat of the exhaust. In September, after modification, the engine was run for the first time on gasoline fuel.

Von Ohain now overtook Whittle and forged ahead. Work moved on to the gasoline-burning HeS 3, which was to be the first proper

Hans Joachim Pabst von Ohain, seen here early in his career. (Wolfgang Muehlbauer)

aero engine, and Heinkel built a small research aircraft in the form of the He 178 (featured in the next chapter) to test it – this aeroplane first flew on 27 August 1939, just four days before the outbreak of World War II. Later came the HeS 8A, which powered the first German jet fighter, the Heinkel He 280 (see Chapter Four), but this aircraft did not enter production. Indeed, none of von Ohain's engines would achieve extended production or operational service because, in the meantime, other design teams in Germany had begun to produce jet powerplants of their own. And some of these proved more successful than von Ohain's creations.

In 1939 Dr.-Ing. Anselm Franz, an Austrian engineer and supercharger expert working for Junkers, took charge of a small team formed specifically to build a jet engine. The result was the Junkers Jumo 004 axial-flow turbojet that was first run on the test bench in October 1940. From July 1942 this would power the twin-engined Messerschmitt Me 262 that became both the fastest jet fighter and the most successful German jet aircraft of World War II.

The prototype Me 262s were at one stage powered by the Bayerische Motoren Werke (BMW) 003. BMW had entered the jet engine field when it bought out Bramo (Brandenburgische Motorenwerke) in 1939. The latter had begun the development of two engines, the 002 that featured a contra-rotating compressor to eliminate torque, and the simpler 003, both under the direction of engineer Hermann Östrich. The 002 was abandoned but, after development problems of its own, the 003 eventually entered production and powered another Heinkel product, the He 162 'people's fighter', as well as the Arado Ar 234C reconnaissance bomber. Incidentally, the German *Reichsluftfahrtministerium* (RLM) or Reich Air Ministry used a '109-' prefix for most German jet and rocket engine projects. So the correct designations for the engines listed here were in fact 109-002, 109-003 and 109-004, while von Ohain's HeS 8 was designated 109-001.

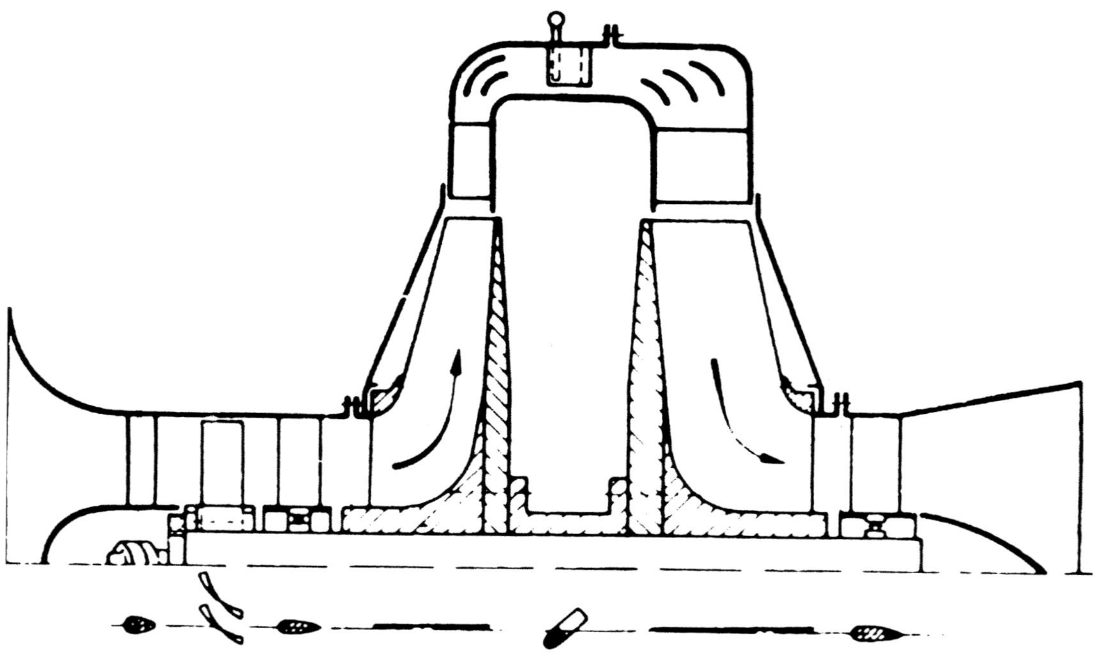

A drawing of the Heinkel-Strahltriebwerk 1 radial turbojet, which was constructed during the second half of 1936 and first run, using hydrogen for combustion, in late March 1937. The unit generated 551lb of thrust. The arrows signify the flow of air through the engine. (Wolfgang Muehlbauer)

In a similar way to Whittle, where the established aero engine manufacturer Rolls-Royce eventually took over from Power Jets, other established German aero engine companies would eventually achieve the biggest successes in the new field of jet propulsion. Nevertheless, it was von Ohain who had paved the way. Although Whittle's work at Lutterworth was unknown to von Ohain, and the latter's at Marienehe was unknown to Whittle (apart from rumours from intelligence sources), the German had been able to study the Englishman's patents. After the war von Ohain went to work in the USA. He also finally met Frank Whittle, and the two would become good friends. Hans von Ohain died in 1998.

ITALY

The engine designer who is probably the least known individual in this story is Secondo Campini of Italy, yet his theories for jet power came to the fore at roughly the same time as Whittle's. In 1930 the journal *L'Aeronautica* published a set of articles written by Campini that examined the possibilities of jet thrust. Then in January 1931 he submitted a proposal to the Italian Air Ministry for an engine that could provide both high speed and performance for an aircraft flying at height. However, the concept contained an important difference to Whittle's ideas. Here, an ordinary radial piston engine was to be used to drive the air compressor before the air passed into a combustion chamber to be mixed with fuel and ignited.

Ingegnere Secondo Campini, inventor of the 'thermojet'. (Archivio Giorgio Apostolo via Marco Comelli)

Campini called his design a 'thermojet' (although the correct term is 'motorjet'), and the gases resulting from the combustion would create high pressure which, like a normal jet engine, would then be ejected out through a rear exhaust. This engine had no turbine, in part to avoid having to develop the new metals that would be needed to make components that could operate at the necessary temperatures and drive a ducted fan. But Campini's thermojet can quite correctly be termed a true jet engine because it used the reactive force of burnt exhaust gases to provide thrust to move an aircraft forward.

In England, Frank Whittle had also looked at the idea of using a piston engine to drive a compressor. However, he had rejected the concept because it was clearly going to be too heavy, and could not provide any clear improvement over the usual combination of a piston engine with a propeller (or 'airscrew' as it was termed at the time). In Campini's system the air would be compressed twice – firstly dynamically through the aircraft's movement relative to it to produce a ram effect as it entered the engine, and then secondly mechanically by the compressor itself. The exhaust velocity and pressure could also be increased or reduced through the use of a movable bullet which would alter the exhaust area.

The Air Ministry rejected this initial proposal, and the next stage in Campini's research saw him become involved with the design not of an aircraft but a boat! In 1931 he and his two brothers moved to Milan and formed the Veivoli E Natanti A Reazione (meaning Jet Aircraft and Boats), or VENAR, company, and in May the *Regia Marina* (Royal Italian Navy) issued them with a research contract to produce and test a marine version of his jet powerplant. One problem was that VENAR had no workshops of its own, so in 1932 Campini established a partnership with Costruzioni Meccaniche Riva of Milan to build a research boat. It was tested in Venice in April 1932 and recorded a speed of 28 knots, which put it on a par with similar boats that used a conventional engine of similar power. However, no production orders were forthcoming.

In 1934 Campini finally secured an order for a research aircraft to demonstrate and assess his jet ideas. Again, because he had no manufacturing facilities of his own, he had to team up with another company, Società Italiana Caproni, to build the experimental aircraft that would evaluate this novel concept (Campini had first approached Caproni back in January 1931, but that move came to nothing). The aircraft appeared as the Campini C.2 and first flew on 27 August 1940 – exactly a year after the He 178 had completed its maiden flight in Germany.

Campini subsequently drew several further aircraft designs, and between 1942 and 1945 worked on a mini-submarine project with Mario de Bernardi that used an oxygen and naphtha-powered jet turbine for underwater operation. At least one prototype was tested in Lake Garda. Campini emigrated to the USA after World War II to work for the Tucker Corporation, and later became involved with the Northrop YB-49 flying wing jet bomber. He died in 1980.

In conclusion, the three key players in this story, Whittle, von Ohain and Campini, all realised that jet propulsion could one day provide powerplants that would take aircraft to speeds and altitudes far higher than had previously been possible. However, Campini's approach to jet propulsion, with a piston engine driving the compressor, proved a dead end. Germany's adoption of the axial jet brought complexity in its designs, but the experience gained would soon be put to good use by the USA, Britain, France and the Soviet Union. Britain's decision to begin with the centrifugal arrangement meant that its first jet engines were simpler, more straightforward to produce and more reliable, but from an early stage it was clear even to Whittle that the axial had to be the way forward for future developments.

In 1945 the state of the art of piston engine development had reached its peak, with the Rolls-Royce Merlin 100 series and Bristol Centaurus propelling types like the de Havilland Hornet and Hawker Sea Fury to record speeds – almost 500mph! However, the presence of a propeller would prevent such aircraft from achieving much higher figures, whereas jet-propulsion engines would eliminate 'airscrews' entirely, thereby opening the path to much greater speeds. By any stretch of the imagination Whittle's and von Ohain's early units were very basic equipment indeed, being raw and totally undeveloped. Yet almost immediately they could take fighter-type aircraft to speeds that virtually matched the very fastest piston-powered machines almost from the word go.

Nobody could have imagined that just five years after the initial flight of the Gloster E.28/39 in May 1941 it would be near universally agreed that the piston engine was 'dead', at least in terms of its use in high performance aeroplanes. But that proved to be the case. When the world's aircraft designers opened their eyes to the potential of the jet, along with new aerodynamic shapes adopted after the war (swept and delta wings), unheard of speeds and performance would soon become available. The aircraft described in the pages that follow took the first steps in European jet-powered flight, and would lead the way to supersonic performance and long-distance operations.

CHAPTER TWO

HEINKEL He 178

This photograph, which was probably taken early in the He 178 V1's career, has been heavily retouched. It is, however, worthy of inclusion because it is one of the few images in existence showing the first aircraft from this angle. (Author's Collection)

The first jet-powered aircraft to take to the air was the He 178, although because of the secrecy surrounding the project and the outbreak of World War II just days after the Heinkel's maiden flight (on 27 August 1939), it would be some time before this fact became known internationally. Constructed purely as a research machine for trials and experiments in jet propulsion, the He 178 would not enjoy the prolonged success achieved by its British equivalent, the Gloster E.28/39 (see Chapter Five), but it was a fascinating and ground-breaking machine nevertheless.

DESIGN PARTNERSHIP

It has been well documented that in Britain Frank Whittle struggled to get sufficient funding and official support for his jet engine projects during the early years of his research work. In Germany, Hans von Ohain was rather more fortunate in that he was employed by a major company, Heinkel, which was in a position to build both his engines and the airframes to use them.

It should be pointed out that, at the same time, there was another young German designer and engineer working for Heinkel who was also looking at powered flight without using a propeller. However, Wernher von Braun's solution, to use a rocket motor as the powerplant, would never find extended use on winged aircraft. But it would find a home in rocket craft and missiles, and von Braun's life's work in this field would eventually culminate in him playing a key role in the US space programme. Back at Heinkel, another purpose-built research

Ernst Heinkel and the Günter twins, Siegfried and Walter, who between them designed both the He 176 and the He 178 research aeroplanes. Walter was killed in a car crash in September 1937. (Wolfgang Muehlbauer)

aircraft in the form of the He 176 was built to test rocket propulsion, and this first flew on 20 June 1939.

Von Ohain's jet development work was carried out in great secrecy inside a special building on Heinkel's Marienehe airfield. The He 178 research aircraft project was started as a private venture, so at this stage the RLM in Berlin was not told anything. Construction of the He 178 began in mid-1938, and its design was largely the work of Hans Regner. The airframe itself was built under the direction of Heinrich Hertel, Heinrich Helmbold and Siegfried Günter. The latter and his twin brother Walter were both aircraft designers and engineers, although Walter was killed in a car accident on 21 September 1937 before construction of the He 178 had begun. He had been involved in the initial design effort, however.

Relatively few photographs of the He 178 appear to have survived, and many of those are of poor quality. Some images have been taken from cine film, which may be the case with this close up. Here, He 178 V1 has no canopy fitted, its main undercarriage well is faired over and the aircraft remains unpainted. The undercarriage itself had a rather old fashioned '1930s' look about it for an aeroplane with such a 'modern' powerplant. (Wolfgang Muehlbauer)

The He 178 was slightly smaller than the Gloster E.28/39, although in general of fairly similar layout. It had shoulder-mounted straight tapered wings of wooden construction that were given a small amount of dihedral. Mounted almost mid-fuselage, these used a profile designed by the Günter brothers and had a thickness/chord ratio of 12 per cent at the root, reducing to 7 per cent at the tips. The conventional flying controls included large inboard trailing-edge flaps, with the ailerons outboard. The tailplane and elevators were also fabricated from wood.

Unlike the wing, the fuselage was a stressed-skin monocoque structure built in duralumin light alloy. Having the wing in its near central position meant that the cockpit was positioned well in advance of the wing leading edges, and its canopy was faired into the dorsal spine. There was a single fuel tank positioned directly behind the cockpit, and then to the rear of this came the jet engine, which had its front end near level with the wing trailing edge. This was fed through a basic, circular pitot-type nose intake, the air then passing along a duct that curved underneath both the cockpit and the fuel tank and in to the engine. The exhaust passed through a long tailpipe (around one third of the He 178's full length) to the tail end of the fuselage. Side intakes had been considered at one stage, although these were rejected in favour of the nose arrangement.

A tailwheel undercarriage was selected with the intention, since it was hoped that speeds in the region of 500mph might be achieved, of having all three wheels capable of retracting into the fuselage. However, for the first flights, the undercarriage was left in the down position and each of the three fuselage openings faired over. The main wheels looked rather large for such a small aeroplane.

A mock-up of the He 178 was inspected by Heinkel and other members of the design team on 28 August 1938. This review resulted in some important modifications being introduced, in particular to the cockpit and canopy, and included the addition of an escape hatch for the pilot fitted into the starboard side of the cockpit.

In the meantime, von Ohain's team had been preparing a flight-standard engine known as the HeS 3, which by March 1938 was

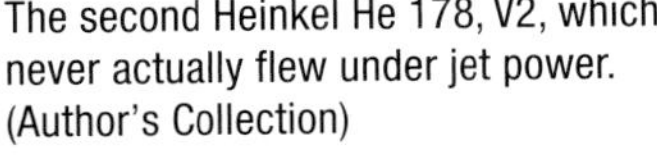

The second Heinkel He 178, V2, which never actually flew under jet power. (Author's Collection)

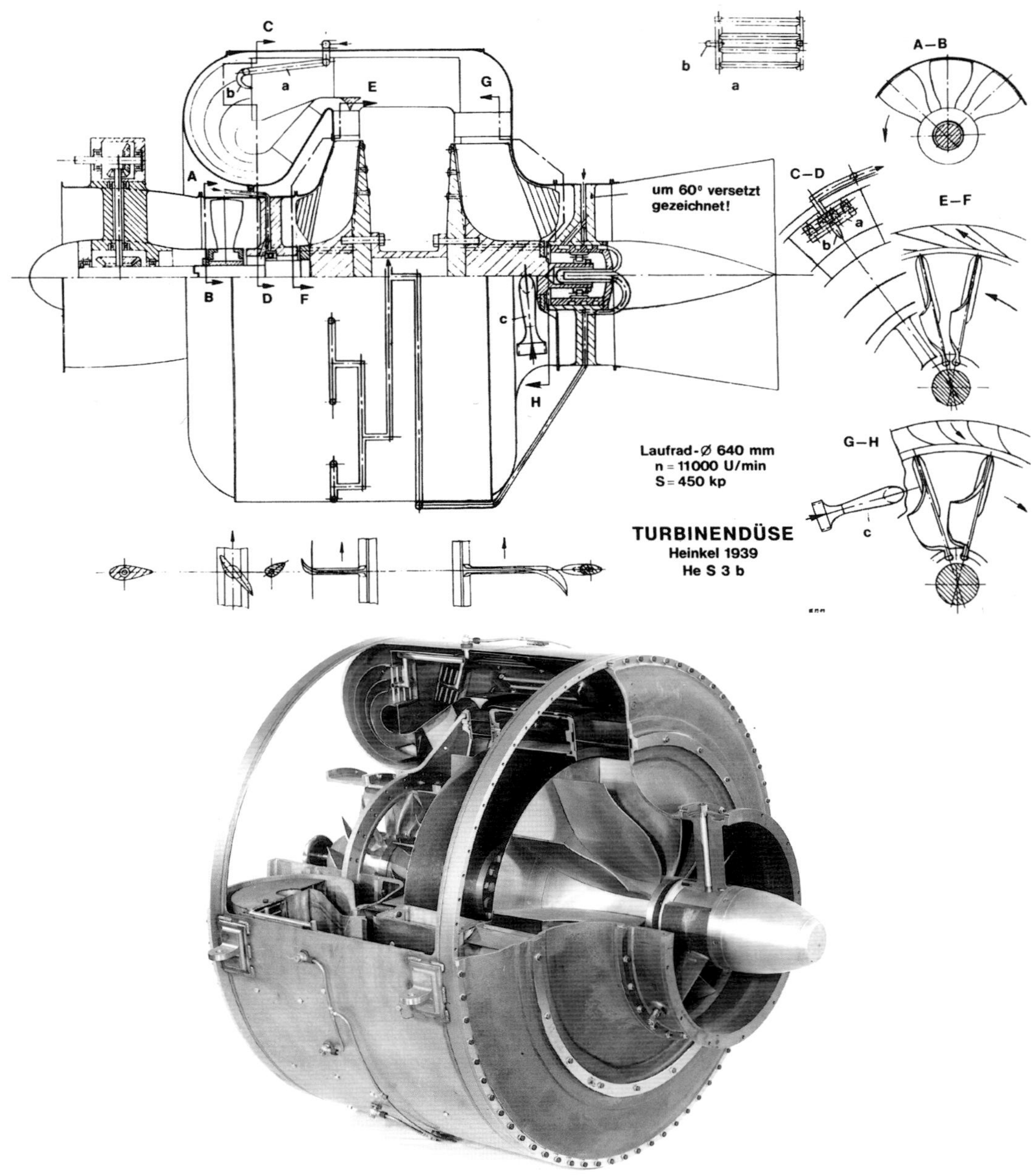

Internal detail drawing and a cutaway display example of the HeS 3B engine that powered the He 178 on its maiden flight. This was the world's first operational turbojet aero engine. (Wolfgang Muehlbauer)

delivering something like 1,100lb of thrust on the test bench. In addition, having used hydrogen to fuel the HeS 1, von Ohain had now switched to petrol – although the HeS 3 still had to be started using hydrogen. Prior to this, there had been the ground-based HeS 2 demonstrator engine that had first been run with hydrogen while bolted to a bench in the spring of 1937. It was switched to petrol from the following September. Although the HeS 2 experienced some problems with its combustor, the engine was responsive to the throttle

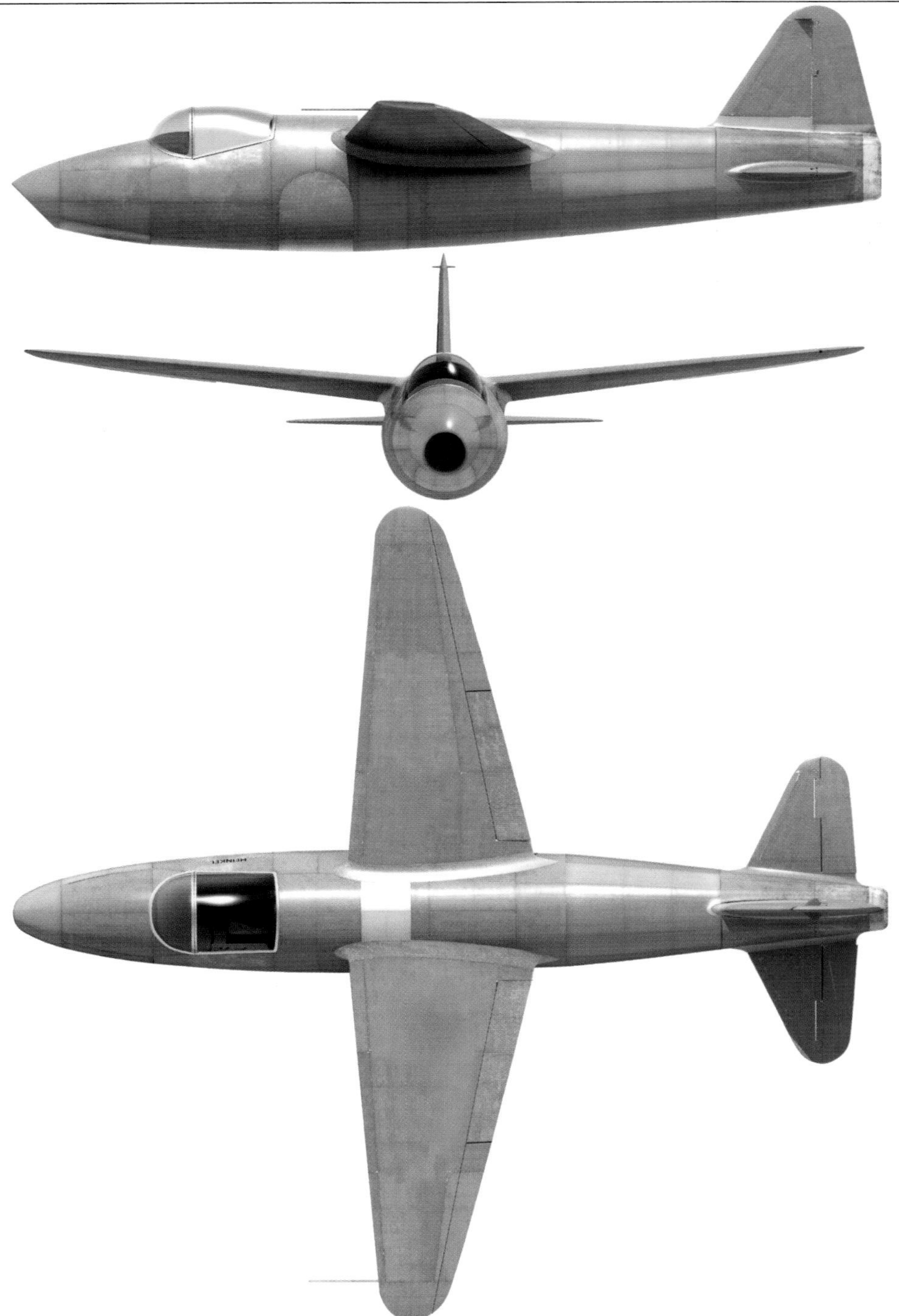

Heinkel He 178

Only two examples of the He 178 were ever built, and both were destroyed during the war. Therefore, with few high-quality photographs existing of either of them (and none in colour), their true appearance is, to some extent, a matter of guesswork. It appears that both aircraft were test-flown in an unpainted state, and were subsequently exhibited in this fashion also. This artwork depicts the V2, which never flew under its own power. The second aircraft introduced wider-span straight tapered wings (the first prototype had elliptical trailing edges) and had a fully retractable undercarriage.

The He 178 V1 seen during taxi trials, or possibly just prior to its first flight. (Wolfgang Muehlbauer)

and provided more vital experience for von Ohain and his team. As a result, in the HeS 3 they had an engine in which the power output could be controlled.

The HeS 3, based largely on the HeS 1, used an axial-flow impeller and large centrifugal compressor. There was an annular reverse flow combustion chamber, but this took only part of the compressed air. The rest of the air would continue rearwards to mix with the combustion gases prior to reaching the turbine, which then drove the compressor. Bench testing of the first HeS 3 began in March 1938, and these runs were disappointing since the unit did not reach the desired thrust rating of about 1,760lb, primarily because the frontal area had been kept as compact as possible through the use of a small compressor and combustor. In due course the HeS 3 design was revised into the larger HeS 3B that eventually developed something approaching 1,100lb of thrust, although early on it struggled to reach 1,000lb. Unfortunately, the He 178's long tail pipe referred to earlier in this chapter would result in around 15 per cent of the thrust achieved in these bench runs being lost once the engine had been installed in the aircraft.

The improvements over the HeS 3 (or now 3A as it is sometimes referred to) involved alterations to the curved guide vanes positioned in between the compressor and combustion chamber and the turbine inlet. In detail the HeS 3B had a 14-blade axial inducer, a 16-blade centrifugal flow compressor and a 12-blade radial inflow turbine. It could run either on petrol (gasoline) or diesel fuel.

Flight testing of an HeS 3A began in May or July 1939 (sources differ), with the unit attached to the underside of Heinkel He 118 D-OVIE – the He 118 had been beaten by the Junkers Ju 87 Stuka in the Luftwaffe's search for a monoplane dive-bomber in the 1930s. This was the first occasion that a flying testbed had been used for a jet engine, and the flight trials were performed in extreme secrecy. The take-off would be made before sunrise (as early as 0400hrs) and the

aircraft would return and land before the factory had opened. Furthermore, D-OVIE would take off and land under propeller-power only. The He 118 proved an ideal airframe for this role because it was a two-seater and it had a long main undercarriage, which meant there was ample space between the ground and fuselage to have the turbojet mounted beneath the aircraft's centre body.

Heinkel test pilot Erich Warsitz undertook all of these flights. Once in the air, the jet was started and observed by company flight engineer Walter Kunzel. Much valuable operating data was gathered from the HeS 3A, which was producing around 840lb of thrust. Describing the first flight of this combination, Ernst Heinkel wrote that after the He 118 had taken off using only its normal piston engine, Walter Kunzel lit the turbine and 'the bluish jet stream shot into the air, and then almost at once the plane was forced up to an enormous speed'. Several flights were completed with the He 118 until, after one landing, the jet engine caught fire and was destroyed. Nevertheless, the HeS 3B was now ready for fitment into the He 178, which itself had by now been completed.

Heinkel test pilot Flugkapitän Erich Warsitz (seen here post-war following five years in a Soviet penal colony in Siberia) had the honour of performing the world's first flights in a rocket-powered aircraft (the He 176) and in a turbojet-powered aircraft (the He 178). (Wolfgang Muehlbauer)

HISTORIC MOMENTS

Prior to its first flight, the He 178 V1 (V for prototype) was despatched to *Erprobungsstelle* Rechlin (E-Stelle – testing station) in June 1939 for a presentation to Adolf Hitler, Generalfeldmarschall Hermann Göring (the Luftwaffe's commander-in-chief) and other officials. Here, Warsitz demonstrated the He 176 rocket prototype in the air before showing the He 178 to the Führer on the ground. On 23 June Warsitz performed the aircraft's first ground runs at the E-Stelle's Rechlin airfield.

As one of the most experienced test pilots in Germany, Flugkapitän Erich Warsitz had been seconded to work with Heinkel and Wernher von Braun on the latter's rocket-powered aircraft in late 1936. As a result, he would become the pilot for the He 176's first flight in June

Although again of poor quality, this photograph of He 178 V1 was snapped just after Flugkapitän Erich Warsitz had taken off from Marienehe on the aircraft's landmark first flight on the morning of 27 August 1939. (Wolfgang Muehlbauer)

This famous photograph shows Ernst Heinkel giving a speech at the celebratory breakfast held in the mess at Marienehe airfield after the world's first successful jet-powered flight. Pilot Erich Warsitz is sitting to the left of Heinkel and engine designer Hans von Ohain is sat on the right. (Wolfgang Muehlbauer)

1939 as well as for the maiden flight of the He 178. This meant Warsitz undertook the world's first flights of aeroplanes using both liquid-fueled rocket power and turbojet power.

At this stage in its development the HeS 3B could not be classed as anything like a reliable engine for flight. Nevertheless, Ernst Heinkel now wanted to have the He 178 in the air as soon as possible. So, back at Marienehe airfield, Warsitz began the aircraft's official taxi trials. On 24 August 1939, during one high-speed taxi test, he made a short straight 'hop' along the runway – the first time that an aircraft had lifted off under turbojet power. The historic first true jet flight followed on the morning of 27 August. At this time the diminutive He 178 had not yet been painted, and it also lacked any official markings.

Before making this famous flight, Warsitz was advised not to fly too fast. With the aircraft then having a fixed undercarriage, the maximum speed during the sortie would have been quite low in any case. With the HeS 3B only permitted to run for six minutes, the pilot had barely managed to complete two large circuits around Marienehe when it was time to land. As Warsitz began his approach to the runway a fuel pump failed. Fortunately, the engine still functioned satisfactorily. Needing to lose height quickly in order to get down as expeditiously as possible, Warsitz performed a sideslip in the He 178 before completing a good landing. A sideslip was a remarkable manoeuvre to execute on the first flight of a brand new and untried aeroplane powered by an equally new and untried powerplant. Warsitz also had trouble seeing the airfield because of glare caused by the low early morning sun reflecting on mist that blanketed the area, the pilot describing this problem as 'terrible'.

This successful flight was a wonderful moment for everyone involved. In the process the HeS 3 had become the world's first operational jet

Ironically, the few photographs that were taken of He 178 V2, the second example to be built, are generally of better quality than for prototype V1. As noted, this He 178 introduced several design changes and it was to be powered by a 1,300lb-thrust HeS 6 engine. (Wolfgang Muehlbauer)

engine to lift an aircraft into the air and to make a complete flight. To celebrate, Warsitz, von Ohain, Heinkel and other members of the team had a champagne breakfast immediately afterwards at Marienehe.

OFFICIAL SUPPORT

There was now a push to get official backing for the project, rather than retaining it as a private venture. In other words, RLM support was now desired and, to this end, on 1 November 1939 the He 178 was demonstrated in flight at Marienehe to Generalluftzeugmeister (Luftwaffe Director-General of Equipment) Generaloberst Ernst Udet, State Secretary of the RLM General Erhard Milch and other high-ranking RLM and Luftwaffe officers. During the He 178's first attempted take-off that day the fuel pump failed. With a serviceable replacement duly fitted, the Heinkel apparently gave an impressive display and performed several high-speed runs at low level.

The flight on 1 November was in fact only the second time the He 178 had been taken aloft because, with the start of World War II, orders had been made to the aviation industry to concentrate from hereon in on production of 'normal' aircraft. Research work was to be slowed down accordingly. The consequent delay had, however, allowed von Ohain to modify his HeS 3B engine into the HeS 6, which was then installed in the He 178 for its second flight. This engine developed 1,300lb of thrust, although the aircraft had in turn seen a weight increase of 132lb – the He 178 now weighed 926lb (the HeS 3B weighed 794lb). As a result, the aircraft's overall performance was quite poor due to the HeS 6's reduced power-to-weight and power-to-diameter ratios. In addition, there were problems with the He 178 itself. The most serious of these was directional instability above certain speeds. The maximum speed flown by the aircraft was about 370mph, and von Ohain felt at the time that the HeS 6 could have enabled the He 178 to reach 435mph had the issues with instability been rectified. Of course the undercarriage was still deployed at this point, and could not be retracted.

The November demonstration received a cool response, and in the short term Heinkel received no orders for further jets, in part because it was

Two photographs showing front and rear angle views of the second He 178. In particular, these images illustrate very well what a simple and straightforward design had been selected for this most important research aircraft – a wise move when the new jet powerplant itself was full of unknowns. (Wolfgang Muehlbauer)

thought that the Luftwaffe could win the war using its existing piston-powered aircraft types. However, a separate problem was that Hans Mauch was the head of the jet engine development section of the *Technische Amt*, the RLM's Technical Department that was in control of all of the industry's research and development programmes. In particular, Mauch saw no place for an airframe company becoming involved with engine development. Fortunately for Heinkel, Mauch left the RLM at the end of 1939 to be replaced by Helmut Schelp, who was more sympathetic to the turbojet programme. As a result, attention at Heinkel turned towards the He 280 fighter and more advanced engines, as described in Chapter Four.

Relatively few details appear to have been recorded in regard to any of the later flying performed by the He 178. However, it is understood that during subsequent test flights made in 1941, with the HeS 6 still in place and now featuring a fully retractable undercarriage, the first

This photograph is reputed to show a mock-up of a planned third He 178 that would have had a longer and more faired canopy and a larger and reshaped fin and rudder. A close examination of the main wheels revealed that they are definitely not rubber tyres, being made instead from wood. (Wolfgang Muehlbauer)

He 178 reached a maximum speed of 435mph. No attempt was made to further develop or make substantial changes to the design of the He 178 itself because, in part, its fuselage-mounted engine had given Heinkel problems that were seen as too difficult to try and overcome at this time. Instead, the He 280 would employ more straightforward wing-mounted engines. In addition, despite its success in powering the He 178 on its first flight, the HeS 3B was in truth underpowered, and von Ohain moved on to a more powerful design altogether in the form of the HeS 8.

A second and slightly enlarged He 178 V2 was also built, but this never flew under jet power. Apparently, it may have flown as an unpowered glider. This aircraft had a larger wing with squared-off wingtips and a gross wing area of 110.65sq.ft. Photographs also show a possible mock-up of an unbuilt third aircraft that featured a longer canopy and an enlarged fin and rudder. The He 178 V1 airframe itself was eventually put on display within the Berlin Aviation Museum. Sadly, however, it was destroyed there in 1943 during an air raid (when the He 176 rocket prototype was also lost). Fortunately, some short film clips of the He 178 still exist to help give us an idea of what this remarkable aeroplane was like. Today, the He 178 is little known to the general public, but it will always be one of aviation history's most important aeroplanes.

HEINKEL He 178	
Type	experimental single-seat jet-powered aircraft
Powerplant	initially 1 x von Ohain/Heinkel HeS 3B turbojet engine giving 1,100lb of thrust. 1 x 1,300lb HeS 6 subsequently installed
Span	23ft 7.5in
Length	24ft 6.5in
Gross Wing Area	97.95sq.ft
All Up Weight	4,273lb (with HeS 3B installed) and 4,405lb (with HeS 6 installed)
Cruising Speed	approximately 360mph at sea level
Maximum Speed	435mph at sea level (with HeS 6 installed)
Ceiling	uncertain
Armament	none fitted

Devoid of any markings or paintwork, one of the Campini C.2s poses for the camera. The tail end of the fuselage had to be of exceptionally large diameter in order to accommodate the Pelton 'bullet' used to regulate the exhaust flow. (Author's Collection)

CHAPTER THREE

CAPRONI-CAMPINI C.2

Over the years there has been some argument as to whether the subject of this chapter, the Italian Caproni-Campini C.2, was a true jet aircraft. In December 1941 Geoffrey Smith reported on the C.2 in *Flight* magazine, summarising the latest news and intelligence about the aircraft and how it appeared to be a major step forward. Later, Smith's writings in general on the new form of jet propulsion were expanded into a book titled *Gas Turbines and Jet Propulsion*, which became probably the first major volume devoted to the subject. Initially published in 1942, this work went through several editions well into the 1950s, and by the sixth edition, in 1955, the C.2 project was being looked upon as something of an also-ran in the jet story. Its powerplant was certainly not a turbojet, but today it is accepted that this Italian effort does fill an important part in the 'jet' story.

The true official designation of this aircraft is C.2 (for Campini 2). Some authors have referred to it as the CC.2 (for Caproni-Campini 2), despite this designation never being used. The two individual prototypes were also known as the N.1 and N.2. This aircraft established several firsts in the jet field. In particular, the C.2 incorporated the first afterburner, and it was the first two-seat jet aeroplane. It was also the first jet-powered type to be revealed to the public – a decision that brought considerable worldwide publicity to Campini and Caproni. In addition, the C.2 was by some margin the largest of the early jet research aircraft, being much bigger than either the Heinkel He 178 or the Gloster E.28/39. Finally, of the three pioneers of jet propulsion, only Secondo Campini would design, build and see fly an aeroplane that was to be powered by a propulsive system of his own creation. Both von Ohain and Whittle had to rely on specialist airframe manufacturers to provide vehicles for their powerplants.

GO-AHEAD

A contract was placed with VENAR (Campini's company) by the *Regia Aeronautica* (Royal Italian Air Force) in February 1934 for a static test airframe, plus a pair of research aircraft fitted with Campini 'powerplants'. With no manufacturing capability of his own, Campini acquired an agreement to use the Aeroplani Caproni factory's facilities at Taliedo, near Milan, in May of that year. However, the requested delivery date of 31 December 1936 would prove impossible to meet, resulting in it being postponed more than once.

Giovanni ('Gianni') Caproni's contribution to the C.2 should not be overlooked. A very forward-looking engineer, he provided both technical and financial support to both Campini and his work for more than ten years. However, this did mean that Campini's designs were frequently classed or termed as Caproni-Campini types, often to the considerable annoyance of the designer. Such was the secrecy associated with this project that the work was undertaken in a secure building known within the factory as the *Centro Sperimentale Campini* (Campini Experimental Centre).

At this early stage Campini had not yet decided what he was going to use to drive the air compressor part of his engine. The choices were to fit either a gas turbine or a conventional reciprocating piston engine. With a delivery deadline only two years away, and lacking suitable materials that could withstand the super-heated environment that existed within a gas turbine, Campini opted to use a conventional engine to drive his compressor.

To this end, in May 1935, Campini put in a request to the *Regia Aeronautica* for the supply of a 12-cylinder Isotta Fraschini Asso XI R piston engine, which he considered would be ideal for the role. However, the development of this particular power unit was still in its relatively early stages, and instead he received an 18-cylinder Isotta Fraschini Asso 750 R. Although less suited to his requirements, this unit did prove good enough for ground testing of the Campini powerplant to go ahead during 1936. To help prove his ideas, Campini also put together a one-third-scale test powerplant that would run during 1935–36, with an electric motor driving the compressor.

Even with the choice of compressor drive established, progress with the airframe design itself slowed to a crawl, thus delaying construction. The starting point had been two paper designs drawn by Campini in 1933 called the CS.500-V and CS.600, which had low- and high-wing positions, respectively. However, apart from their wing arrangements, these were quite similar designs, with each sporting a nose air intake, tubular fuselage, two-seat cockpit positioned to the rear of the wings, a fixed undercarriage and the compressor engine installed in front of the cockpit. They were then blended together to form the design chosen for manufacture, which from the start was a large aircraft that, for a time, retained the fixed faired-in undercarriage. Scale models were

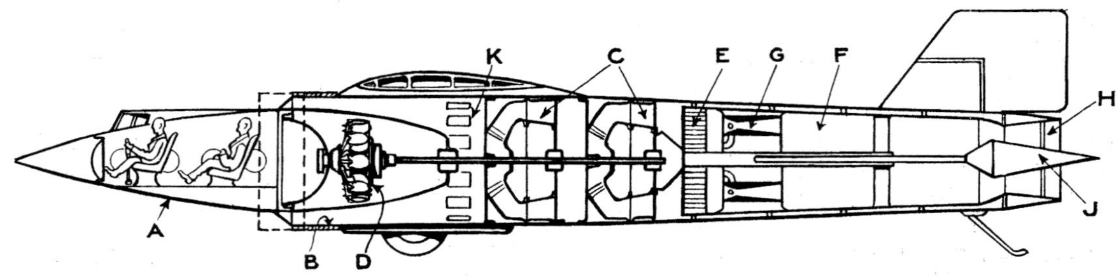

A. Ovoid cabin.
B. Enshrouding cylinder.
C. Two-stage centrifugal compressor.
D. Radial engine.
E. Rectifier-radiator.
F. Combustion space.
G. Annular mixing channel.
H. Discharge nozzle.
J. Nozzle control bullet.
K. Controlled lateral orifices.

An early Campini design for a 'motorjet' aircraft that would have used a piston engine to drive the compressor, but which never went beyond the drawing board. It was in fact published as a patent in America and is dated 17 December 1935. The aircraft would have had a pressurised cockpit, and the project was described (rather optimistically) as being capable of flight at either subsonic or supersonic speeds. (Author's Collection)

tunnel tested both in the Caproni wind tunnel and also at the Guidonia test centre near Rome.

On 6 December 1936, Campini reported that the fuselage had been completed (in fact the two machines were built side-by-side), but he would need a further six months to get the first aircraft ready. On 27 April 1937, the official engine tests were started, and witnessed by a military board headed by Generale Ingegnere (General Engineer) Enrico Bonessa. However, the results were disappointing, as they fell short of the original calculations. A thrust of 1,433lbs was recorded, which was about 7 per cent down on the estimates. With the burners running, only 1,610lbs of thrust was produced, some 81 per cent of the calculated figure. The Campini burners also consumed rather more fuel than had been expected – in fact five times as much!

Continued slow progress and rising costs meant that in July 1937 Campini had to ask for additional funding from Government Undersecretary Giuseppe Valle to cover a 30 per cent rise in expenditure. He also requested a delivery deadline extension. The *Regia Aeronautica* eventually agreed, and in December 1937 the contract was amended, with the delivery date now stated as 31 December 1938. Later, this would be moved back again to 31 October 1939, and in the end the C.2 took until the summer of 1940 to be ready to fly.

COMPLEX DESIGN

The C.2 had a substantially cylindrical parallel-sided fuselage – in essence a tube slightly tapered at its ends, and which had a maximum external diameter of 62in and internal diameter between approximately 55in and 56in. It was formed in four sections, namely:

1. The air intake nosing.
2. A forward portion which contained the compressor.
3. The main centre fuselage section that housed (from front to rear) a circular coolant radiator, the piston engine and the 'afterburning' equipment. A British report from RAE Farnborough stated that 'there was considerable internal obstruction in this section of the fuselage due to the protrusion of the cabin into the duct, and by internal struts and supports for the adjustable bullet equipment'.

The two Campini C.2 fuselages seen under construction inside the Caproni factory. Their all-metal structure was, at the time, unusual practice for the Italian aircraft industry. The first aeroplane, N.1 (construction number NC4849), is on the left. (Archivio Giorgio Apostolo via Marco Comelli)

4. The tail portion housing the combustion chamber and the adjustable Pelton cone or 'bullet' that controlled the final exit orifice area.

The semi-monocoque fuselage was built around circular formers and stringers and was double-skinned, the inner skin corrugated, the outer skin smooth (the object of the inner skin was to reduce losses in thrust brought about when the airflow was obstructed by the framing and other excrescences). Most of the fuselage was constructed in aluminium light alloy, but the tail section had to be lined with steel to protect it from the considerable heat generated by the exhaust gases. The cockpit was to be pressurised (this facility was never installed) and both seats had individual rearward-sliding canopies and dual controls, which meant that the aircraft could be flown from either position. The cockpit instrumentation was normal and, in fact, pretty basic, although additional instruments were introduced specifically to operate the aircraft's new type of powerplant.

The thick chord low-position wing was elliptical in shape and built in one piece. It had twin spars and four-section trailing edge flaps. The ribs used channel-section frames and had holes cut in them to save weight, there were large wing root fillets and the entire fuel load was carried inside the wing. Two spars were also used to build the tailplane, which was situated at the base of the vertical fin, and both rudder and elevators were horn balanced. The empennage went through a lot of alterations during the C.2's development and also after flight testing had started, with several changes in shape. A tailwheel 'taildragger' landing gear was fitted which, by the time the two aircraft were built, had been made entirely retractable, the main legs folding outwards into the wing.

The Campini C.2's powerplant was based around a 900hp 12-cylinder vee liquid-cooled Isotta Fraschini Asso XI L.121 RC40 reciprocating (piston) engine that drove a three-stage ducted fan or compressor. The compressor had three fixed stages each with 15 vanes (their pitch could be adjusted hydraulically), and three rotating stages each with 16 blades – the pitch of these blades could only be altered on the ground before take-off. An annular radiator was positioned to the rear of the compressor to cool the RC40 unit.

Air was provided by the nose intake, and this would undergo an initial compression just from the forward motion of the aircraft, before being compressed further by means of the compressor itself. The air would then exhaust into the fuselage at the rear of the fan, before reaching the 'afterburning' equipment in the form of a vapourising burner provided in the rear fuselage near the tailwheel. Here, fuel could be added to the compressed air and then ignited through the supplementary burners to provide additional heat. The heated and expanding gases were finally directed along the tailpipe to produce the jet thrust that would then propel the aircraft forward. The afterburner was actually a circular grid supporting a set of fuel injectors and flameholders, and this system would be turned on primarily for high-speed flight at high altitudes. In other words, in 'normal' flight conditions the powerplant could also run 'cold', that is without the burners operating. To enable the afterburner to be tested on the ground, the flameholder grid could be exposed by the removal of the entire rear fuselage section.

The cross-sectional area of the final exit orifice was controlled by an adjustable Pelton bullet (a cone on the end of a shaft that was arranged to slide longitudinally to vary the cross-sectional area of the jet orifice). The bullet's movement, and hence the speed of the propulsive jet, came under the control of the pilot. A Pelton 'bullet', or more accurately 'needle', is a plug normally used in Pelton turbines to regulate the intake of fluid.

As mentioned earlier, ground tests had been conducted with an 18-cylinder engine. The redesign necessary to fit the 12-cylinder RC40 flight engine, two examples (one for each airframe) of which were supplied to Campini at Taliedo in March 1940, brought serious increases in weight to the point that the original 1934 estimate for an airframe empty weight of 2,645lbs eventually rose to around 7,720lb for the completed aeroplane. Campini's Motorista (engine specialist or technician) Casalini was responsible for the redesign. Apparently Campini never called this powerplant a turbojet – as noted in Chapter One, the term he used was 'motoreattore' or 'motorjet'.

Besides the C.2's correct designation, the two airframes individually received multiple labels. As well as the codes N.1 and N.2, they would both carry military serials, MM.487 and MM.488, respectively. These were numbers that came within a block of serials reserved specially by the *Regia Aeronautica* for experimental aircraft and prototypes. The Caproni factory also allotted its own construction numbers, NC4849 and NC4850, respectively.

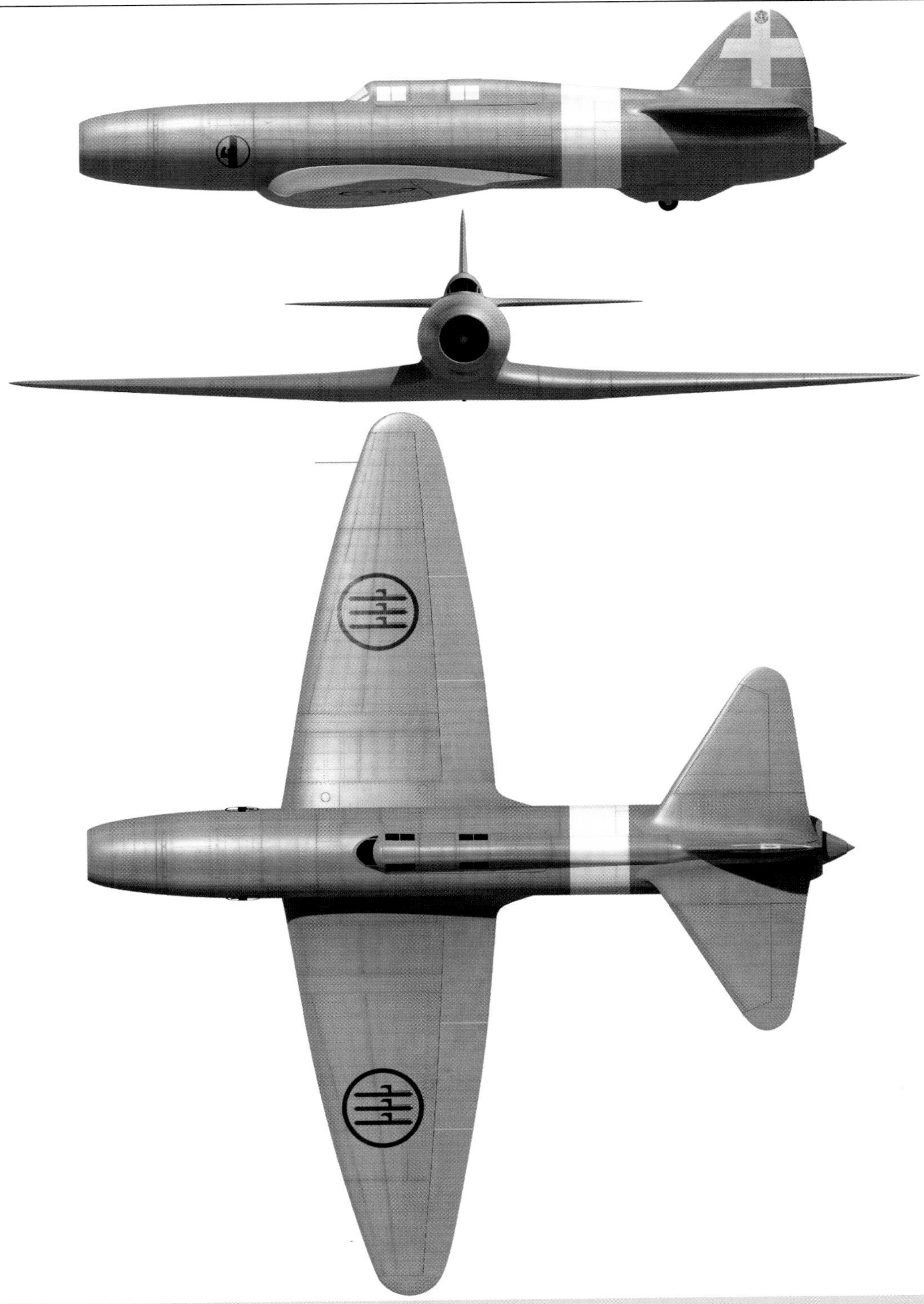

X PLANES

CAPRONI-CAMPINI C.2

This three-view of Caproni-Campini C.2 MM.487 shows the aircraft in the markings it wore when delivered to the DSSE at Guidonia on 30 November 1941 following its famous Milan to Rome flight.

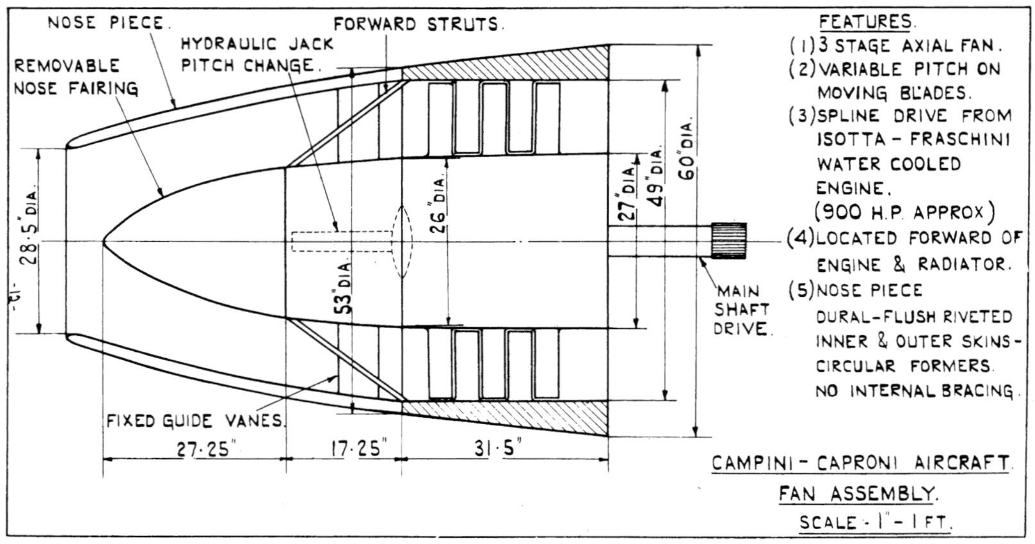

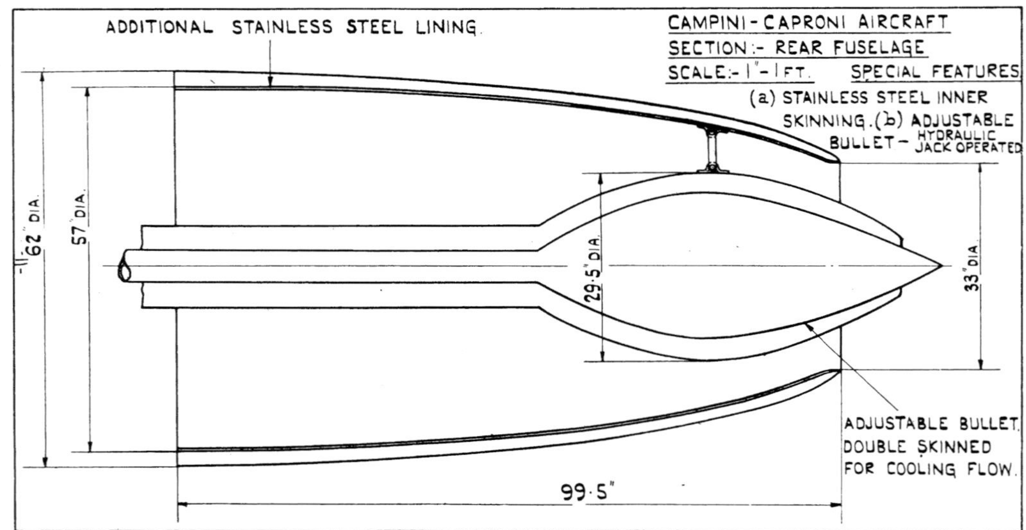

Drawings taken from, it is thought, Sqn Ldr F. E. Pickles's report on the C.2. They show, firstly, the C.2's fan assembly, and then the rear fuselage and exhaust with its Pelton 'bullet'. (Author's Collection)

READY TO FLY

Ground test runs with the flight engine in place, in fact in airframe N.2 and with Casalini in the cockpit, were started on 28 June 1940, and these indicated that the compressor generated about 1,555lb of thrust. In the meantime, the famous aviator Colonello Mario de Bernardi had come on board as the nominated C.2 test pilot. In 1915 de Bernardi had been the first Italian credited with downing an enemy aircraft in aerial combat during World War I, and in November 1926 he had won the Schneider Trophy race for Italy flying a Macchi M.39 racing seaplane.

On 26 July 1940, Campini told the Constructions Directorate that both machines were now ready for their contractual flight trials. On

8 August, after the aircraft had been brought the short distance from Taliedo by road, de Bernardi took N.2 on its first taxi trial at Linate airfield in Milan. These runs lasted an hour and resulted in adjustments being made to the undercarriage. N.2 undertook these initial trials after the decision had been made to use N.1 for the type's fatigue and airworthiness tests, which were mandatory for an airworthiness certificate to be awarded to the C.2. A second set of taxi trials was made on 27 August 1940, a year to the day after the Heinkel He 178 had made the world's first flight by a turbojet-powered aeroplane. However, during these trials, at 1935hrs de Bernardi took off to complete a maiden flight that lasted ten minutes. On this trip he did not light the afterburner, but nevertheless this was Italy's first 'jet' flight.

A run of 2,300ft was required for take-off, but because of the lack of power the aircraft's rate of climb was found to be quite poor (different sources quote 138ft per minute and 300ft per minute). Thus it was the second C.2 that became airborne first, and the event was reported in *Il Popolo d'Italia* newspaper, which of course alerted the rest of the world to the C.2's existence. Despite the short duration of this sortie, de Bernardi was able to establish that the C.2 had a tendency to dive, and that the elevators were over-sensitive. As a result, along with other aerodynamic alterations, the tailplane was modified by reducing the angle of incidence. These changes were also made to N.1.

The first official contract flight took place on 16 September 1940, but here again the afterburner was not lit. Further engine ground runs took place with Casalini in the cockpit between 17 and 19 September, then a few days later de Bernardi broke his heel when he tumbled into the lift-shaft of his hotel. Since Campini would not trust another pilot in his aeroplane, there was to be no further flying for some time, although from October through to April 1941 Casalini continued to carry out ground runs. Flight testing resumed on 11 April 1941, after de Bernardi had been cleared to fly again. This flight, another ten-minute sortie, was made with aircraft N.1 (its first flight), which had replaced N.2 in the flying programme. It was also the first occasion that the burners were operated in flight. On 5 May the very first take-off using the afterburner was made, which reduced the take-off run and improved the rate of climb to 1,100ft per minute. However, during the course of this sortie, the RC40 piston unit experienced vibration and backfiring, and later in the summer it was replaced.

Flight six was made on 1 June 1941, and was simply a short display sortie with the afterburners running to show off the aircraft to *Regia Aeronautica* Undersecretary Francesco Pricolo, who was visiting the factory at Taliedo. Flight seven came on 7 July, and was a much longer affair, lasting a full hour and with the afterburner lit for 15 minutes, during which it functioned satisfactorily. By then it had been discovered that the burners were prone to continued localized overheating, and this necessitated yet more ground testing from May to July 1941. The N.1's first flights with the replacement piston engine in place were made on 19 and 20 October 1941, but as before, there was vibration and backfiring when unburnt fuel ignited in the exhaust pipe. Once again the piston engine had to be removed and overhauled.

Testing the afterburner on the first C.2 at Taliedo just prior to its famous flight to the *Regia Aeronautica*'s research and test establishment at Guidonia. A crane was used to remove the completely detachable and quite large rear fuselage section. Ingegnere Secondo Campini is standing behind the jet flame, while his chief engine specialist Casalini looks back from the cockpit. (Archivio Giorgio Apostolo via Marco Comelli)

With the switch to test flying using aircraft N.1, N.2 (later serialled MM.488) became almost totally inactive. In fact it appears to have flown just once more, on 31 August 1941, on what was apparently an official contract flight. This was certainly the last recorded flight of N.2, and it took place in front of the acceptance board led by Generale Ingegnere Bonessa.

IN THE PUBLIC EYE

Perhaps the biggest highlight in the C.2's relatively brief career was a landmark and quite famous flight made from Linate to Guidonia, just to the east of Rome. The new venue was the *Regia Aeronautica*'s test and research airfield, where the official C.2 evaluation trials were to be conducted. This facility, the *Direzione Superiore degli Studi e delle Esperienze* (DSSE), established in 1935, was responsible for assessing all of Italy's new military aircraft, as well as undertaking basic aeronautical research. It was something of a combination of Britain's RAE Farnborough and the Aeroplane and Armament Experimental Establishment (A&AEE) at Boscombe Down, in Wiltshire, and was Italy's primary aeronautical experimental establishment. Its facilities included six wind tunnels (one of which was a small supersonic tunnel), a seaplane testing department and various laboratories.

Prior to this cross-country flight, de Bernardi made several preparatory sorties in the silver-painted airframe N.1 to check that the aircraft was ready for this important journey. On 19 October and 6 November 1941, he flew to altitudes of 4,920ft and 8,200ft, respectively, and then on 7 November completed an endurance flight of one hour. In addition, on 5 November, de Bernardi took aloft the C.2's first passenger, Ingegnere Giovanni Pedace, and on 29 November Commendatore Guasti (who was Caproni's nephew). These gentlemen duly became the world's first jet passengers.

The very long nose of the Campini, or perhaps more relevant the mid-fuselage cockpit position, must have made the pilot's view forward pretty poor. (Author's Collection)

On 30 November 1941, de Bernardi and Pedace took off from Linate in N.1 at 1447hrs to make the delivery flight to the DSSE at Guidonia. They successfully landed there at 1658hrs, having covered the 298 miles in two hours and 11.5 minutes at an average speed of just 135mph. However, this was not the full story – the distance covered was in fact rather more because adverse weather had made a diversion to Pisa necessary, although in the end de Bernardi did not have to land the aeroplane at this intermediate stop and they were able to continue on to Guidonia. Only the forward engine was used for the flight, and de Bernardi even let Pedace fly N.1 for about 15 minutes.

This relatively straightforward ferry flight in fact created quite a sensation, particularly in the Italian press, because it had been turned into a full-scale media event with worldwide publicity. Timekeepers were even brought in to measure the flight duration accurately. The flight was also logged as the world's first official *Fédération Aéronautique Internationale* record for any jet aircraft type. However, Campini stressed that high speed had not been the objective of this particular flight, which gave the public their first view of his aeroplane.

On 5 December 1941, de Bernardi was asked by Prime Minister Benito Mussolini to fly the aircraft over the centre of Rome. The following day Mussolini inspected N.1 at Guidonia, and de Bernardi performed another demonstration flight. A few days later display flights were undertaken for Generales Eraldo Ilari and Amedeo Mecozzi of the *Regia Aeronautica* and Ingegnere Zappa. During the official evaluation the N.1 would also be used for ceremonial flights and displays for overseas visitors (which must have caused some disruption to the programme).

The DSSE Guidonia report covering the C.2's official trials, dated October 1942, did not show much enthusiasm for the aircraft. Immediately after the arrival of N.1 at Guidonia, work began on installing the necessary measuring and recording instruments, which was completed by 12 December. Eighteen days later, N.1 was given its

The first aircraft, N.1, performs a low pass over the airport at Pisa during its transfer flight to Guidonia. Although the C.2 was not a pure turbojet aircraft, and nor did it possess anything like an impressive performance, it is easy to forget that this was still the first Italian aircraft to fly without a propeller. Little wonder so much publicity was given to it by the nation's press and publicity machine. (Archivio Giorgio Apostolo via Marco Comelli)

new military serial MM.487, and the official trials began on 7 January 1942. Unfortunately, the first attempt to calibrate the airspeed recorder in flight was unsuccessful on account of bad weather conditions, and then for the second attempt on 13 January 1942, the flight had to be abandoned due to the engine cooling system suffering a leak. Indeed, throughout the trials MM.487 was to be plagued by minor problems that saw it grounded several times.

Flight tests at low altitude began on 6 April 1942 in connection with the calibration of the airspeed indicator. Three days later a climb and altitude performance test was attempted, but owing to unsatisfactory weather conditions just one speed test (at an altitude of 3,280ft) was completed. During the course of this flight the radiator overheated, which meant that pilot de Bernardi needed to land quickly. However, a failure in the landing gear hydraulic circuit jammed the port leg wheel in the lowered position, while the starboard leg remained retracted.

N.1 is made ready for its flight from Taliedo to Guidonia. On the left in the white overcoat is the aircraft's pilot, Colonello Mario de Bernardi, while the tall individual in the centre of the group wearing the hat is Ingegnere Secondo Campini. Co-pilot Ingegnere Giovanni Pedace, in flying overalls and a jacket (with a fur collar) stands behind Campini. Note the wing and cockpit access ladders. (Archivio Giorgio Apostolo via Marco Comelli)

Thus de Bernardi had to land on one wheel, but thanks to his skill the damage suffered by MM.487 was restricted to just the tail wheel and its adjacent fairing, the rear fuselage and a wing rib in the undercarriage recess. As a result, the hydraulic circuit of the undercarriage retraction gear was altered. This flight had also shown up irregular engine running and excessively high temperatures in the engine and oil cooling areas, which also required modifications.

Flight testing resumed on 10 June 1942 with a sortie during which the pilot reached a speed of 181mph. Another climbing test was undertaken two days later, and this again proved unsatisfactory because the fairing flap of MM.487's starboard gear leg stayed open during the entire flight. This test was repeated after the undercarriage had been overhauled, but the results were no better because the engine was unable to maintain the stipulated charging pressure when the aircraft was flying above 2,300ft. Continuing leakage in the engine cooling system also necessitated further repairs, which lasted until 23 July. Then in the ensuing ground tests a tube fracture in the valve block of the undercarriage retraction mechanism made the starboard leg fail. Fortunately, this caused only light damage, and the repairs were completed by 28 July.

The aeroplane remained grounded for some time until two climb and altitude performance assessment flights were made on 26 and 27 August 1942. The first of these was another failure, this time due to the irregular firing of the engine burners, but the results obtained on the second flight were good enough to enable the manufacturers to bring the test programme to an end. In view of the difficulties revealed in flying sorties with aeroplanes of this type, it was considered that further flights were unlikely to produce any improvement in the performance

This air-to-air photograph of N.1 was taken during the aircraft's flight from Taliedo to Guidonia in late November 1941. Note that both cockpits are open – de Bernardi never took the aircraft above an altitude of 1,640ft, and most of the trip was flown in the 985–1,640ft range. Eventually, the C.2 was to have been fitted with a pressurised cockpit, but this work was never carried out. (Archivio Giorgio Apostolo via Marco Comelli)

figures obtained so far (a decision with which the DSSE concurred). Hence, the 27 August trip was the C.2's last (known) flight.

The results of the test flights were inconclusive owing to the difficulty of obtaining and maintaining completely horizontal and steady flight in the C.2. The Guidonia DSSE report concluded with the following statements:

'1. The aircraft is sluggish in taking off, even with the burners on. Considering the low all-up weight, a better performance could be expected from an aeroplane fitted with a normal airscrew.

'2. The fuel consumption of the burners is exorbitant compared with the slight advantage derived from their use, whether in horizontal or climbing flight.

'3. The additional effect of the burners is small and was difficult to determine owing to failures in different parts of the aeroplane.

'4. The trials have demonstrated the practical possibility of flight by other than normal airscrew propulsion, but, considering the very low speeds attained in the test flights, the advantages of jet propulsion are hardly shown in a favourable light. These performances could be improved upon by any aeroplane of the same dimensions and weight fitted with a normal airscrew.

'5. In view of the novelty of this form of aircraft propulsion, the tests just completed can hardly be termed conclusive. It remains to be seen whether, under present conditions, further tests would be capable of giving better results. The manufacturers do not appear to have carried out the necessary number of preliminary tests for a complete development of the design.'

Having just landed N.1 at Guidonia, de Bernardi has carefully extricated himself from the cockpit and climbed out onto the wing root, while Pedace stands in the cockpit with a briefcase or bag. The latter contained mail with a special cancellation to mark what was also the first ever airmail flight using a 'jet' aircraft – another landmark for the C.2. (Archivio Giorgio Apostolo via Marco Comelli)

Although the DSSE completed only a small number of flights in MM.487, these were sufficient to reveal that the machine possessed a rather poor performance. It was soon clear that the experimental C.2 could not be considered as a successful design in any way, and the *Regia Aeronautica* could see no future in the project. From this point MM.487 was held inactive at Guidonia.

POST-WAR EXAMINATION

The National Archives in Kew, in the London Borough of Richmond-upon-Thames, hold many fascinating wartime files compiled from the various pieces of intelligence sent back to Britain from different sources in the frontline. Amongst them are documents revealing the existence of the first German and Italian jet aircraft, as well as detailed descriptions of these projects. Within the files for the C.2 is a report by British specialist Sqn Ldr F. E. Pickles of the British Ministry of Aircraft Production, who inspected the damaged N.1/MM.487 at Guidonia after German forces had been driven out of the area.

After the research establishment at Guidonia finished with MM.487, it had been placed in storage in a hangar. Pickles, who was on attachment to Mediterranean Allied Air Forces Intelligence specifically to assess jet aircraft and engine developments found in Italy, examined N.1 and its powerplant on 18 June 1944. He found it partially dismantled, with the fuselage and intake cowling lightly damaged by shrapnel, and with the nose section, the three-stage fan (compressor), the one-piece wing and the tail empennage (in three sections) all separated from the fuselage.

Although the wing skin had been cut apart above the wheel wells to remove the tyres, and all of the glazing had been smashed, the airframe was mostly complete. Pickles also spoke to 'experts' and eyewitnesses who had seen the aircraft operating and were now prisoners of war – these included test pilot de Bernardi.

One intelligence report stated that the C.2 had three conditions of flight:

'(a) Low-speed flight – the total propulsive thrust was produced by the air compressor driven by the engine and, therefore, the performance of the aircraft was comparable with that of an ordinary aircraft type with a reciprocating engine and a propeller. Additional energy in the form of heat was added to the air both by supercharger compression and from the cooling system of the engine.

'(b) High-speed flight – injectors were switched on and further heat energy was provided by direct combustion of the fuel.

'(c) Flight with the [jet] engine stopped – the aircraft could continue to fly, but the fuel consumption would increase enormously.'

De Bernardi added that he had flown the C.2 at speeds varying between 93mph and 373mph (the latter figure seems rather unlikely), and that the aircraft was very easy to fly. On another occasion, he stated that the concept of an aircraft that did not have a propeller was looked upon with skepticism both in the Caproni factory and elsewhere – each of the jet designers in this book had their disbelievers! It was also reported that the aircraft which de Bernardi had flown was 'not a prototype, but embodied certain features which were considered necessary in the case of an experimental aircraft, the performance of which could not be predicted. The wing area of the experimental model was much larger than would have been chosen for the prototype of a production series.' During the flying conducted at Guidonia MM.487 could often be seen over Rome, and it was conspicuous for the continuous siren-like sound that it emitted. Sadly, Pickles also reported that the Guidonia Experimental Establishment was now 'completely demolished'.

Pickles next made arrangements to have MM.487 despatched to the RAE at Farnborough, where it arrived in October 1944. Following further detailed examination the airframe sections were stored in January 1946, pending a possible transfer to a museum. However, another inspection in November 1947 found that the metal structure had now become seriously corroded, so the pieces were despatched to RAF Newton, in Nottinghamshire, from where they went in 1949 to a local scrapyard. Parts of the fuselage and the fan section were last spotted in this yard in 1951.

In contrast, MM.488, having remained at Linate, survived the war in fine condition. In April 1952 the *Aeronautica Militare* (Italian Air Force) recovered the aircraft and put it on static display at numerous exhibitions and airshows. Since 1977 it has been on show in the *Museo Storico dell'Aeronautica Militare* (Italian Air Force Museum) at Vigna di Valle, having been repainted to represent how MM.487 looked during its famous Milan to Rome (Linate to Guidonia) flight. In addition, the static test fuselage of 1937 also survived the war and was apparently still at Taliedo in 1950. Since the mid-1950s this has been on display in the

A colour view of N.1 (or rather MM.487, as it had now become) at Guidonia on 29 January 1942, and with a Piaggio P.108 bomber parked behind. Both aircraft have just been inspected by an official delegation from the *Magyar Légierő* (Hungarian Air Force). There are few photographs showing the C.2s with their military serial numbers, which were applied in small dark letters and numbers on the lower rear fuselage underneath the forward part of the tailplane. (Archivio Giorgio Apostolo via Marco Comelli)

Museo Nazionale della Scienza e della Tecnologia (National Museum of Science and Technology) 'Leonardo da Vinci' in Milan.

It would appear that more photographs of the C.2 were published in the aviation press during the early to mid-war years than any other jet-powered type, in part of course because of the Milan to Rome flight. The Italian press published a lot of images, but their accompanying descriptions were purely of a non-technical nature and intended only for propaganda purposes. In Britain, of course, the Gloster E.28/39 was on the Secret List, and so was not revealed to the public for some time, while illustrations of the Heinkel prototypes outside Germany were usually artworks prepared from reports provided by Allied aircrew or intelligence contacts on the ground who had encountered the jets. As a consequence, the C.2 gained more early publicity worldwide than perhaps it should have done.

In the end it was probably a mistake to order two examples of the C.2 well before its propulsion system had reached any form of maturity. As a consequence, the programme was seriously delayed and the resulting additional costs became substantial. As a pure research tool for this new form of powerplant, several sources have indicated that the C.2 was just too big and too complicated – factors that resulted in a performance which in the end was really quite poor. In addition, when considered alongside the other engines and aircraft in this book, the C.2's ducted-fan system was somewhat crude and had no chance of competing with the turbojet engine – a fact that many had become aware of by 1943. The Campini C.2 was too inefficient and had too much drag and too little thrust to provide it with a worthwhile performance. Nevertheless, it was the world's second jet aircraft to fly.

In July 1944 Caproni drew a sleek twin-engined bomber design that looked a little like the British de Havilland Mosquito but fitted

MM.488 has been on display within the *Museo Storico dell'Aeronautica Militare* at Vigna di Valle since 1977, the aircraft having been repainted to represent how MM.487 looked during its famous Milan to Rome (Linate to Guidonia) flight. (Courtesy of Francesco Farina via Marco Comelli)

with twin fins. This was to have been powered by two Campini units mounted in wing nacelles but, alas, it was never built. The first true jet aircraft design to come out of Italy's aircraft industry was drawn in about 1945, and this too did not progress to hardware. The first purely Italian jet aircraft to fly was the Fiat G.80 military trainer, which made its maiden flight in December 1951.

Data quoted for the C.2 varies considerably with different sources, in part because of the aircraft's frequent modifications. However, the figures presented here relate to the aircraft in its final form as flown:

CAPRONI-CAMPINI C.2	
Type	experimental two-seat jet-powered aircraft
Powerplant	1 x 900hp Isotta Fraschini Asso XI RC40 (L.121RC40) air-cooled piston engine driving a 1,550lb thrust compressor with burner
Span	48ft 0in
Length	39ft 8in
Gross Wing Area	392.69sq.ft
Loaded Wight	9,720lb
Maximum Speed	204mph without afterburner, 233mph at 9,800ft with burner engaged (the Guidonia report gave 202mph without burner and 223mph with burner, both at 9,840ft)
Ceiling	13,300ft – the climb to this height took 53 minutes. With the burner running the C.2 could reach 3,280ft in about nine minutes
Armament	none fitted

CHAPTER FOUR

HEINKEL He 280

For gliding trials, the He 280 V1 was fitted with ballasted dummy engine nacelles to provide data as accurate as possible for the aircraft's low-speed flight characteristics. (Wolfgang Muehlbauer)

The Heinkel He 280 jet fighter is the only aircraft in this book to have been designed from the beginning for a production run and frontline service. Consequently, more than two examples were built and flown. However, another of Germany's leading aircraft designers, Messerschmitt with its Me 262, also produced a jet fighter that in some respects proved superior to the He 280. As a result, Heinkel's product never reached a squadron and was never involved in any action.

STUDY AND STRUCTURE

Heinkel designer Robert Lusser began serious work on a jet fighter prior to the He 178 demonstrating the feasibility of turbojet propulsion. Before the end of 1938 Messerschmitt had received a formal order to develop a fighter with a turbojet powerplant (which became the Me 262), and in response Ernst Heinkel at Rostock-Marienehe decided to go ahead with a jet fighter of his own powered by an engine designed and built by his company. As noted in Chapter Two, the jet engine development section of the *Technische Amt* was not keen on the idea of an airframe company producing aero engines as well. However, in October 1939, the official designation He 280 was applied to the fighter and, though it was still a private venture project, Heinkel committed his company to building three prototypes (in-house, the He 280 had previously been referred to as the He 180) and their engines.

As if having a turbojet powerplant was not enough in the way of advanced features, the He 280 also introduced another radical

Robert Lusser was the designer of the He 280. Aside from his many years with Heinkel, he also worked for Messerschmitt (specifically on the Bf 108 and Bf 110) and Fieseler (where he was involved in the design of the Fi 103, which evolved into the V1). (Wolfgang Muehlbauer)

innovation in the form of an ejector seat for the pilot. This was probably the first such seat developed anywhere in the world, and it was operated using compressed air. The seat was also designed and built in-house by Heinkel during the period June to November 1940, and when actuated, it could propel a pilot of average weight some 20ft from his cockpit.

The He 280 also had a fully-retractable, hydraulically-operated, tricycle (nosewheel) undercarriage that allowed the fighter to taxi with its axis horizontal. This was important because the first examples of the Me 262 were fitted with a conventional tailwheel undercarriage. This meant that their fuselages were angled downwards towards the rear, as were their turbojet thrust lines. Messerschmitt's test pilots soon experienced problems as a result of the engines pointing at the ground during the initial stages of their take off roll, the motors struggling to develop enough thrust to raise the tail of the Me 262.

The He 280 used an all-metal stressed-skin structure throughout. Its low- to mid-position wing had a straight untapered leading edge, with a trailing edge that was semi-elliptical, while outboard of the engines there was a small amount of dihedral. The HeS 8 jets themselves were suspended in nacelles beneath the forward and centre wing spars and just outboard of the attachment points of the inward-retracting main undercarriage legs – their thrust lines and jet efflux passed underneath

Examples of the Heinkel HeS 8 (left) and Junkers Jumo 004 engines placed side-by-side for comparison. Unlike the HeS 8, the Jumo 004 proved to be a successful first-generation jet engine. (Wolfgang Muehlbauer)

the raised tailplane. Plain trailing edge flaps were fitted in two sections, one on either side of the jet pipes, and ailerons were positioned further outboard. On retraction, the main undercarriage wheels were stowed inside fuselage wells to the rear of the cockpit, with the legs in the wing. The high-mounted tailplane, lifted above the fuselage by a shallow pylon, had a fin and rudder on each tip and a one-piece elevator in between.

There was a slender monocoque fuselage of oval section that helped give the He 280 a very clean aerodynamic shape. The cockpit was positioned immediately ahead of the wing leading edge and the original design had provision to install cabin pressurisation. Three MG 151/20 20mm cannon formed a powerful armament, with the guns being grouped together symmetrically in the nose over the front wheel well.

A He 111H-3 was adapted as a test bed for the HeS 8 engine, with the power unit mounted on a pair of pylons underneath the former bomber's fuselage. These pylons or struts were telescopic, allowing them to be extended in flight to ensure that the jet engine would be functioning in relatively undisturbed air. (Wolfgang Muehlbauer)

A primary requirement for any fighter-type aircraft was to have sufficient room to house a suitable armament. The research types described in the previous chapters did not have this need, and so each of them used a nose air intake (this was probably a coincidence, but it helped to keep things simple). By reason of the wing position of its jet units, the He 280 was able to have its guns concentrated in the nose, and that had the further advantage of providing enough space to carry plenty of ammunition.

ENGINE DEVELOPMENTS

While Lusser was designing the He 280, von Ohain had continued with his engine development work for Heinkel, which included the HeS 8 earmarked for the company's pioneering jet fighter. The politics behind the various German jet engine programmes during World War II would have a substantial influence on the fortunes of the He 280, so this aspect needs to be reviewed here in reasonable detail.

Following on from the HeS 6 flown in the He 178, the HeS 8 introduced a straight-through flow combustion layout within a more compact size than the HeS 6. In fact, the effort given to the HeS 8 had resulted in a unit that was slimmer than all previous von Ohain engines, and it had a combined 14-blade impeller and 19-blade compressor, both manufactured in aluminium. Its 14-blade turbine, however, was made of steel and, since it had no form of cooling, the turbine would have 'burnt out' in a very short time. The HeS 8 was expected to provide a static thrust of 1,540lb, and it would eventually become the company's first jet engine to receive RLM financial support under

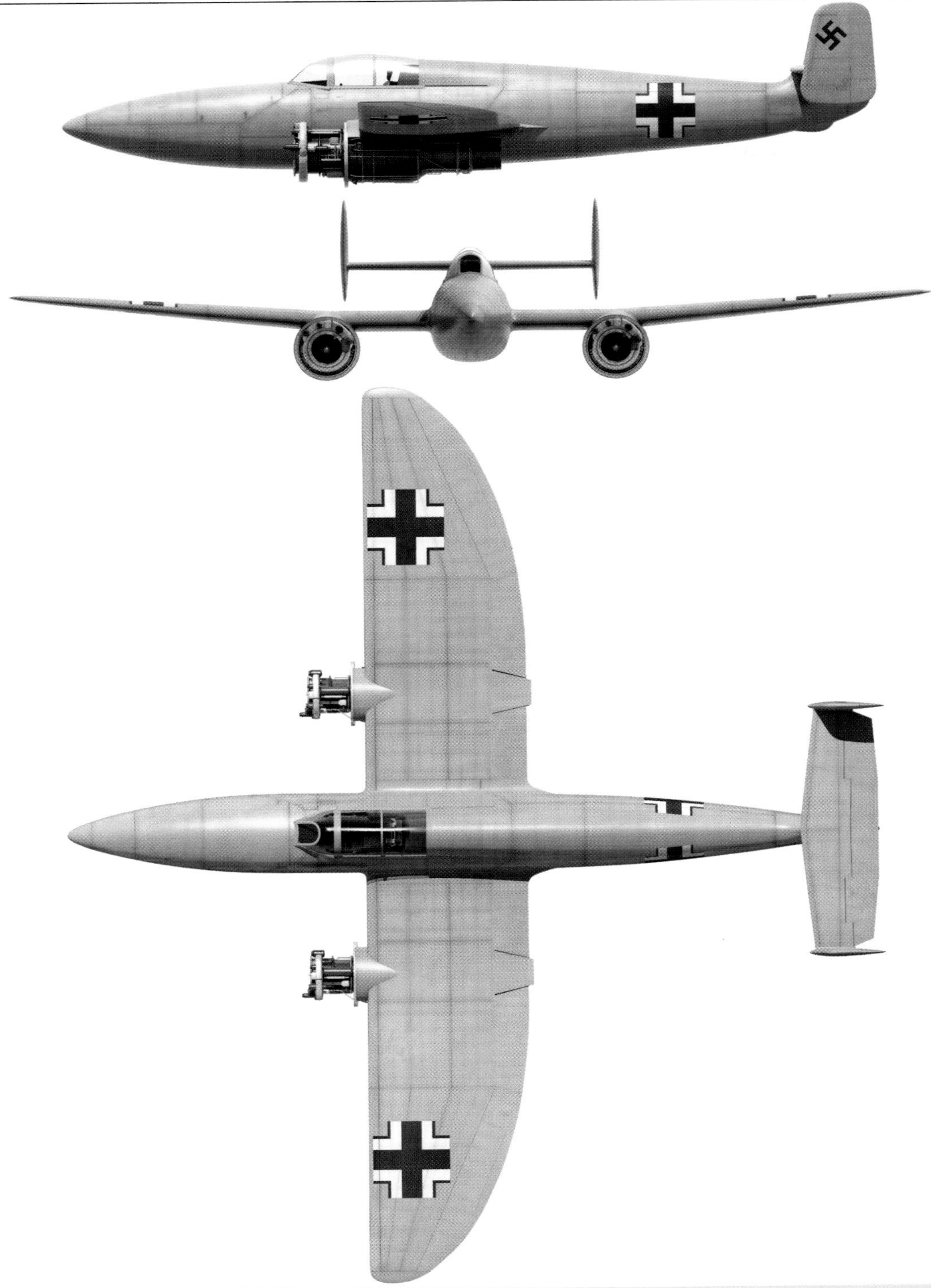

HEINKEL He 280

Devoid of the four-letter codes that would subsequently adorn either side of its fuselage, He 280 V2 became the world's first jet fighter to take to the air under its own power on 30 March 1941 at Marienehe. As this three-view clearly shows, the aircraft flew without cowlings fitted to either of its HeS 8 engines.

The first prototype He 280 V1 is towed down the snowy runway at Rechlin in late 1940 by a He 111 tug during one of its trial sorties as a glider. Test pilots Bader and Schäfer were both involved in this critical phase of the flying programme. (Wolfgang Muehlbauer)

the designation 109-001 (in Luftwaffe service it would have been the Heinkel 001). However, Heinkel also had a new axial jet engine design in the pipeline called the HeS 30.

The 1,760lb thrust HeS 30 (known officially as the 109-006) was designed by engineer Adolf Müller, who had recently joined Heinkel at Marienehe from the Magdeburg factory operated by Junkers. The HeS 30 had a five-stage compressor, ten individual combustion chambers and a single-stage turbine. Müller and von Ohain worked on their new engines almost in parallel, and both types were scheduled to power examples of the He 280. By late 1939 Heinkel's jet engines were also facing strong competition from Junkers with its Jumo 109-004 and BMW (Bramo) with the longer-term and more advanced 109-003 programme, both of which employed axial compressors.

On 9 April 1941, with the He 280 now undergoing flight testing, Heinkel was cleared to buy the Hirth Motoren GmbH of Stuttgart-Zuffenhausen, a company that had specialized in producing piston engines and turbo-superchargers. Müller and his HeS 30 were now transferred to Stuttgart, which allowed von Ohain to access all of the Heinkel facilities at Rostock for both the HeS 8 and for a new, more advanced, engine called the HeS 11 (officially the 109-011). Progress with the HeS 8 was, however, disappointing, and by the end of 1941 the Junkers 109-004 had overtaken it in terms of development.

Consequently, since the HeS 30 engine was showing rather more promise (with a superior thrust-to-weight ratio), and because von Ohain had completed the HeS 11's design, the HeS 8 was abandoned in 1943 – it would be test flown in the He 280 V3 prototype as part of the HeS 11's development work. Müller left Heinkel during late spring 1942 after a violent disagreement, by which point his HeS 30 was coming along well, developing 1,980lb of thrust. There was a strong desire to see this project retained, therefore. In the autumn of 1942, however, Helmut Schelp at the *Technische Amt* (who felt that the BMW 003 and Junkers Jumo 004 were already 'good enough')

Fritz Schäfer returns to Marienehe at the end of He 280 V2's very brief maiden flight under its own power on 30 March 1941, the aircraft only having enough fuel in its tanks to complete a single circuit of the airfield. Nevertheless, this flight made aviation history, for it was the first time a turbojet-powered aircraft that had been designed as a fighter from the start had taken to the air. This angle shows particularly well how exposed the Heinkel jet engines were when their cowlings were removed – considered at the time a necessary step because in those early days unburnt fuel could collect at the bottom of the cowlings and create a serious fire hazard. (Wolfgang Muehlbauer)

ordered that Heinkel should now concentrate on the HeS 11 and cease development of the HeS 30.

This directive coincided with the HeS 11 (which was now described by Schelp as a 'Class II' engine since it was offering in the region of 2,870lb of thrust) also being transferred to Stuttgart. So, in a short space of time, the He 280 had lost both of its intended Heinkel-designed engines. A replacement was now urgently needed.

The first HeS 8 had been ready for bench testing as early as April 1940, although the design had many shortcomings resulting in it being beset by numerous problems. Indeed, it was still only producing 1,210lb of thrust in early 1942. Von Ohain made several major design changes on a number of prototype engines that eventually pushed the thrust up to about 1,320lb. An improved version designated the HeS 8B (the original duly became the HeS 8A) was produced in 1942, this powerplant introducing a new compressor and turbine. Examples of the HeS 8 were test flown underneath the fuselage of a Heinkel He 111 test-bed aircraft.

FIRST FLIGHTS

The first prototype He 280 V1, which received the *Stammkennzeichen* (RLM aircraft designation or code) DL+AS, had been completed by September 1940. However, its HeS 8 engines were not yet ready, so the initial trials were conducted with the aircraft as an unpowered glider. As such, it had dummy nacelles (ballasted pods to represent the weight of engines and fuel) attached in place of the engines themselves. Subsequent research flying would confirm that the He 280 was quite a good glider, despite its advanced design. The first gliding flight was

Ernst Heinkel poses for an informal photograph with the first two pilots to fly the He 280 jet fighter, Paul Bader (in uniform) and Fritz Schäfer (right). This shot was taken following Bader's demonstration flight in front of several high-ranking Luftwaffe officers and RLM officials on 5 April 1941. (Wolfgang Muehlbauer)

made on 22 September with Luftwaffe test pilot Fliegerstabsing Paul Bader in the cockpit. The fighter was towed aloft by a He 111 bomber, with the He 280 casting off from its tug at a height of 13,120ft. During his descent Bader recorded a gliding speed of 174mph, with the flight lasting just six minutes. Nine further gliding flights were completed by the end of October 1940, with 280mph recorded during flight six and 317mph on flight nine.

From November the trials were shared between Bader and Heinkel test pilot Fritz Schäfer, the latter making his first sortie in V1 on 20 November. The early gliding flights were conducted from the *Erprobungsstelle* Rechlin (one of the Luftwaffe's flight test establishments for new aircraft) until the programme was moved to Marienehe in

This grainy photograph shows one of the He 280 prototypes in flight with two HeS 8A engines installed. The image's lack of clarity prevents any code letters showing up on the rear fuselage, making the exact identity of the aircraft difficult to confirm. It is believed to be V3, pictured during its first flight on 5 July 1942. (Wolfgang Muehlbauer)

Prototype V3 is lined up with the runway as the pilot comes in to land at the end of its first flight on 5 July 1942. (Wolfgang Muehlbauer)

November. Some 41 flights had been recorded by 17 March 1941, and these were sufficient to establish that V1's handling qualities at lower speeds were satisfactory.

Two HeS 8 engines were finally available in March 1941, although they were only providing 1,100lb of thrust rather than the intended 1,540lb. The He 280 finally took to the air under its own power on 30 March 1941, and prototype V2 (code GJ+CA) performed this historic flight rather than V1. The pilot for the occasion was Fritz Schäfer, who wrote about the experience in 1958:

'I had joined the Heinkel company at the beginning of 1941 after working for some time as a civilian engineering test pilot at the Luftwaffe's armament test centre at Tarnewitz. Erich Warsitz [the He 178 pilot] had left Heinkel during the early months of the war to take up a post at the experimental centre at Peenemünde.

'This first flight was obviously to be a matter of some delicacy. The HeS 8 turbojets had never been tested in the air and had in fact only just completed one hour of bench running. The cowlings had been left off as it had been discovered that quantities of gasoline tended to collect in the bottoms of these while the turbojets were running, presenting a very serious fire hazard. The 280 V1 was not entirely new to me of course, as I had flown the machine as a glider on several occasions in order to gather data on the general characteristics of the aircraft in the air.

'My main thought was to conserve just enough fuel in my tanks for a single flight around Marienehe airfield. The greatest difficulty was the fact that the primitive turbojets had no automatic regulating devices, the throttle cock being the only means of controlling the thrust of the powerplants. I had to watch the engine rev counters extremely closely as it was dangerous to exceed 13,000rpm even by 200 or 300 revs,

V3 again, this time on 8 February 1943 after a turbine blade had broken away and pilot Fritz Schäfer had had to land with the undercarriage up. By all accounts this aircraft flew again only three days later. (Wolfgang Muehlbauer)

and below this figure the turbojets did not develop sufficient thrust to enable me to take-off.

'I was airborne without much runway to spare. I climbed to 275m [902ft] and throttled back to 12,000rpm, beginning my circuit. There was no time to raise the undercarriage because the red warning lights were blinking, indicating that I was already flying on reserve fuel. I turned over Warnow, made a fairly fast approach and landed comfortably on the runway with 40 degrees of flap. The world's first jet fighter had flown successfully!'

The He 280's fuselage tank held 1,070 litres of fuel, but for its maiden flight only 400 litres had been provided – a figure considered sufficient for the planned sortie. The flight between take-off and landing lasted a mere three minutes and, under instruction, Schäfer never intended to retract V2's undercarriage. He later reported that, without any device to synchronize the revs of the two engines, the only real problem had been balancing the thrust using the throttles. Fully loaded, this He 280 weighed 9,435lb, and in due course it would reach a maximum speed of 485mph at 19,685ft.

A demonstration flight before high-ranking Luftwaffe and RLM officials took place just days later on 5 April. This time the He 280 had its turbojets fully cowled and Paul Bader was the pilot. Among those present to watch the demonstration were Generaloberst Ernst Udet, Helmut Schelp and General-Ingenieurs Roluf Lucht and Wolfram Eisenlohr, who were responsible for the Luftwaffe's airframe and engine departments, respectively, at the *Technische Amt*. The demonstration proceeded without incident and the onlookers were highly enthusiastic. A small celebration was subsequently held in the factory, during which (according to Fritz Schäfer's account) Udet told Ernst Heinkel that 'If the British saw a few such machines over the Channel they would start scrapping the whole of their own programme!' This demonstration flight convinced Eisenlohr of the soundness of Germany's new jet

This superb nose-angle close-up of He 280 V3 provides excellent details of the rather heavy-looking undercarriage and the front of the engine nacelles, one covered and the other with its front cowling removed. Note the wing dihedral outboard of the engine nacelles. (Wolfgang Muehlbauer)

aircraft types and helped to secure some official backing for Heinkel's jet programme. However, Udet's enthusiasm proved to be short-term, and later he would question the need for this new type of warplane.

PROTOTYPES, POLITICS AND TEST BEDS

Prototypes V1, V2 and V3 (code GJ+CB) all undertook their maiden flights powered by the HeS 8 engine. V1 made its first powered flight at the start of April 1941, with V3 following a full 15 months later. After all of this time the HeS 8 was still only producing a little over 1,320lb of thrust. Just one more He 280, prototype V5 (CJ+CD), would make a maiden flight with the HeS 8 installed on 26 July 1943. A modest increase in engine thrust, however, did allow V3 to reach a speed of 497mph (it was hoped that the HeS 8B could push the He 280's maximum speed up to 578mph).

NEXT PAGES

Had Heinkel succeeded in convincing the Luftwaffe to commit to series production of the He 280, this 'what if' scene may have been the reality of air combat over the Third Reich as early as the spring of 1944. Here, He 280As of 9./JG 3 engage B-17Gs of the 94th Bombardment Group at high altitude over western Germany, the fighters' trio of nose-mounted MG 151/20 20mm cannon taking their toll on the bombers. With a top speed in excess of 500mph at altitudes above 35,000ft, the He 280A would have been more than a handful for P-38J Lightning, P-47D Thunderbolt and P-51B/C Mustang pilots charged with escorting the four-engined 'heavies' on raids against German targets.

A
A

A
775
A

Then on 8 February 1943 V3 lost a turbine blade in its starboard engine shortly after take-off, which in turn produced some intense vibration and a sheet of flame that streamed well back from the nacelle. Fritz Schäfer was at the controls of the aircraft at the time, and, having cut the damaged engine and then tried to maintain a straight course in spite of the asymmetric thrust from the port-side HeS 8, he was able to make a forced landing with the undercarriage retracted. Incredibly, even with near-full fuel tanks, the resulting damage was light and, repairs and with a replacement engine in place, V3 flew again just three days later (this aircraft was eventually captured by the Allies at Schwechat, south-east of Vienna, in May 1945).

Apart from the lack of engine thrust, as the He 280 was reaching higher speeds there was now evidence that it was experiencing tail flutter – the oscillation was only encountered at certain speeds. There was also some tendency towards directional snaking when flying under specific conditions. Overall, however, both pilots (Schäfer and Bader) stressed that the fighter's handling characteristics were, in general, excellent.

In the early summer of 1942, following RAF raids on Rostock in April and May, the decision was taken to move the He 280 development flying from Marienehe to Schwechat. By this time the HeS 8's future prospects were fading, and consideration was being given to fitting different engines. The first He 280 to receive alternative power units was the prototype V2, which first flew with 1,850lb thrust Junkers Jumo 004B-0s as replacements for its HeS 8s on 16 March 1943, the installation having been undertaken at Schwechat. Fritz Schäfer made the first flight, but when preparing to land he found that the flaps would not deploy. He was forced to make a high-speed landing and, having reached the end of the completed runway (which at the time was being extended), had the undercarriage torn off as he passed over a trench. Nevertheless, V2 was repaired and duly flew again.

With the Jumo 004s in place, the prototype was able to record a level speed of 497mph at 13,120ft. However, the airframe and, especially, the tail assembly lacked sufficient strength to cope with the vibration caused by these engines, which meant that the pilot had to reduce the speed. V2's career with its 004s ended on 26 June 1943 when the aircraft was destroyed after an undercarriage failure on landing – at that time it was being used for research into snaking at high speeds.

On 27 March 1943, Generalfeldmarschall Erhard Milch, Air Inspector General of the RLM, officially cancelled both the He 280 jet fighter (in preference to the Me 262, which was by now in direct competition with it for a production order) and its HeS 8A engine. All further development would be halted, although the nine prototypes on order were to be finished and small numbers of HeS 8s would be produced to prevent any untoward hold-ups in the planned flight test programme. The He 280s would then be found alternative roles as flying test-beds to benefit the other concurrent German development programmes for new jet engines. The He 280's airframe design had revealed weaknesses (severe airframe tail flutter had now been experienced when flying above 500mph), and it had inadequate fuel

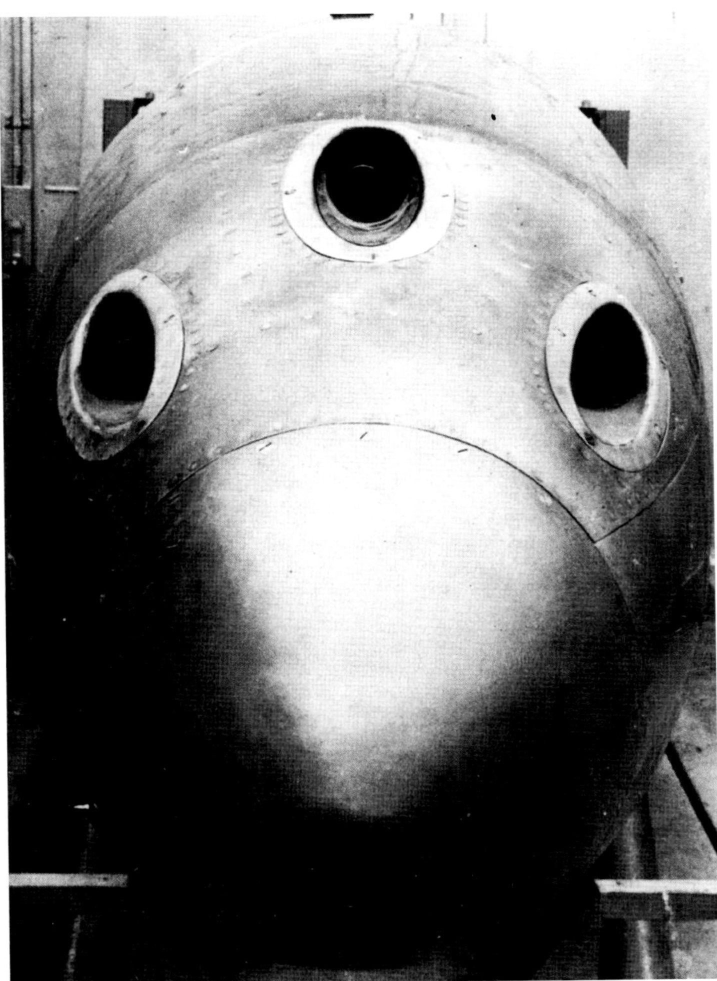

Nose armament of three MG 151/20 20mm cannon was installed in two He 280 prototypes, V5 and V6. (Wolfgang Muehlbauer)

tankage. The latter was the result of gross underestimates in the level of fuel consumption of the early jet units, which in turn left the Heinkel fighter with a woefully short range.

The Me 262 made its first flight on 18 April 1941, albeit with piston engines. Its first flight under jet power alone came on 18 July 1942, and the judgement was that the Messerschmitt was the superior fighter. In 1941 there was an attempt to fly an Me 262 with two HeS 8 engines installed, but this was a rather heavier aircraft than the He 280 and these relatively low-power units proved incapable of getting the Messerschmitt fighter off the ground.

Ernst Heinkel had tried very hard to get his fighter accepted for service orders despite the RLM's growing disinterest. For example, a series of mock combats was arranged against a Focke-Wulf Fw 190A provided by the Arado factory at Warnemünde, which was close by. The Fw 190 was one of the very best piston fighters then in service, but the He 280 conquered its adversary and apparently proved superior in a number of respects. As a direct result of these trials, the RLM

Prototype V7 joined the DFS at Ainring to undertake aerodynamic research trials as a glider. The engines were removed but, unlike V1 during its early gliding flights, there were no dummy engine nacelles, and that gave the aircraft a rather odd appearance. Other alterations included the addition of the long pitot tube seen over the nose. In this form the aircraft made its maiden flight on 19 April 1943. (Wolfgang Muehlbauer)

instructed Heinkel to make preparations to build 13 He 280A-0 pre-production aeroplanes (the engines to be fitted in these machines were not specified).

Also, prior to the final abandonment, Heinkel had firmed up his design for the production version of his fighter, which was to have been designated He 280B-1 – this was proposed officially to the RLM in early 1943. The problems surrounding the aircraft's poor range were to be dealt with via a fuselage extension of 2.6ft, increasing the jet's fuel capacity by a third. The tail unit and empennage were to be improved through the introduction of a single fin and rudder to replace the twin fin arrangement, the nose gun armament was to be increased to six MG 151/20 20mm cannon and a bomb rack would be fitted underneath the fuselage. An increase in wing area was necessary to compensate for the rise in weight. With the loss of the HeS 8, the He 280B-1 was to be powered by a pair of 1,984lb thrust Junkers Jumo 004B engines. Although some loss of performance was predicted, the aircraft's estimated maximum speed was still 547mph.

Prior to this proposal the RLM had considered that series production of the He 280 was unlikely, one reason being Heinkel's other wartime commitments which were considerable. However, negotiations now began towards the manufacture of 300 He 280B-1s, with the work to be undertaken by Siebel Flugzeugwerke Halle on a sub-contract. New estimates for the production of the Jumo 004B engine made soon afterwards indicated that this plan was entirely unrealistic however, so these orders were never placed.

Despite Generalfeldmarschall Milch having effectively killed off the He 280 in March 1943, within the Heinkel establishment there was still great interest, with further prototypes nearing completion. Schäfer reported how the tempo of the flight test programme had risen, and on the thrill of reaching 'fantastic' speeds in the region of 500mph in level flight (which in 1943 could indeed be considered as a high speed).

Prototype V4 (code GJ+CC) was the first He 280 designed to have different engines fitted from the start. It had two 1,650lb thrust BMW 003s, and as such it first flew on 31 August 1943, the flight trials

In terms of reaching frontline operational service with the Luftwaffe, the He 280 lost out to the Messerschmitt Me 262 powered by Jumo 004 engines. This aircraft was arguably the most advanced jet fighter to see combat during World War II. (Author's Collection)

being conducted at Schwechat. V4 apparently had Jumo 004s installed late in September 1944, and it was eventually Struck off Charge at Hörsching, in Austria, in October of that year. As previously noted, prototype V5 was completed with Heinkel-Hirth HeS 8 units, but on 15 September 1943 it made another first flight, now with BMW 003s installed. This aircraft and prototype V6 (NU+EA) were both used for armament trials, having had the specified three 20mm cannon mounted in the upper nose. At one stage V5 was considered to be the pre-series prototype for the planned production He 280. V6 first flew with BMW 003s on 26 July 1943.

The final He 280 prototype to be completed was V9 (code NU+ED), which first flew with BMW 003A-1 engines in its nacelles on 31 August 1943. In this form V9 was used at Rechlin as a 003A-1 test bed until May 1944. V5, V6 and V9 all went into store at the end of their flying careers. V7 and V8 are discussed shortly, but consideration was given for constructing three more examples, V10, V11, and V12, which apparently would have been full prototypes for the production He 280B-1.

FURTHER TRIALS

Several of the He 280 prototypes continued to fill important roles as engine test beds or in other trials programmes. Aerodynamically, the type was well suited for the work, and having underwing-mounted engines made their replacement by different types of power unit a relatively straightforward process. It could be argued that all of the flight testing conducted using the various He 280 airframes was of a trials nature.

Both the first and fourth examples, V1 and V4, had Argus 109-014 pulsejet engines fitted under their wings in place of the normal turbojets, and as such were tested at Rechlin. This simple engine was to be used to power the Fi 103 (V1) pilotless missile or flying bomb, and the He 280 trials formed part of the weapon's development programme. It was quickly found that the Argus units generated substantial acoustic vibration that could badly damage the airframe. They also failed to generate sufficient thrust to enable the fighter to become airborne independently, so it had to be towed aloft by a pair of Messerschmitt Bf 110 piston-engined fighters.

The initial trial using prototype V1 fitted with four pulsejets came to an abrupt end on 13 January 1942 when the towing cables and the aircraft became iced-up following a heavy snow shower, to the point where the former could only be released by the Bf 110s. Helmut Schenk, the test pilot for the Argus development programme, could not communicate with his tug pilots, and so he had no option but to abandon the He 280 since he could not land with long pieces of cable hanging off the nose after the Bf 110s had released the two lines from their end. In doing so he made the world's first ever ejection seat bail out. The plan had been to release the He 280 at an altitude of 7,875ft and then light the pulsejets. Prototype V4 fitted with six pulsejets served as the replacement, which enabled the Argus programme to be continued more successfully.

Prototypes V7 (NU+EB) and V8 (NU+EC) undertook trials of a different kind, being involved in aerodynamic test flights. V7 was taken to the *Deutsche Forschungsanstalt für Segelflug* (DFS – German Institute for Glider Research) at Ainring. Here, in readiness for a full aerodynamic research programme, which was in fact undertaken at Hörsching, near Linz, the engines were removed and instruments for recording the aircraft's attitude in flight, and cameras to film threads attached to the wings, were added. In addition, a long pitot tube was attached over the nose. V7 made its maiden flight without any engines on 19 April 1943.

The designated test pilot for this programme was Wilhelm Mohr, and its objective was to investigate ways of increasing the Mach numbers and Reynolds numbers of aircraft wings, and to examine other problems associated with flight at high speeds. Over an 18-month period 115 gliding flights were completed with the aircraft. Standard procedure was to use the DFS's He 111H-6 to tow V7 aloft, and most of these sorties were performed with the wing kept close to the stall. In addition, on about 50 occasions the aircraft was put into a spin, but its stalling and spinning behaviour never reached a critical state. The highest altitude to which V7 was taken by the He 111 was 23,000ft, and the high-level sorties also enabled some high-speed dives to be made, during which the aircraft experienced no unpleasant characteristics. One source states that the V7 reached a speed of 578mph during a dive, and it also recorded 310mph on several occasions. Later in its career this He 280 was allocated the civil registration D-IEXM.

He 280 V8 was selected to investigate the aerodynamic characteristics of the 'butterfly' or V-tail, wind tunnel and other testing having

The only Heinkel jet fighter type to reach production was the He 162, which entered service too late in the war to have any major impact. The photograph is thought to show the first prototype V1, which made its maiden flight from Schwechat on 6 December 1944. (Author's Collection)

indicated that this arrangement might offer some advantages over conventional tailplanes when flying at high speeds. Fitted with Jumo 004 turbojets from the start of its flying career, V8 first flew on 19 July 1943, and in horizontal flight would achieve a maximum speed of 435mph (on 30 August 1943). The V-tail was introduced soon afterwards, and the resulting data later proved valuable when the tailplane for Heinkel's He 162 fighter was being designed.

As noted, He 280 V6 first flew in July 1943, and then during 1944 the aircraft was refitted with a single fin-and-rudder assembly in readiness for comparison flights tests with the V-tail V7. It first flew in this new configuration at the start of 1945, but was then lost in a crash near Berlin during its initial high-speed trial.

UNDER SCRUTINY

It is interesting to quote some wartime Allied Intelligence reports that show how, in British eyes, the He 280 programme had been progressing. The first reports of work by Heinkel on jet-propelled aircraft types date back to April 1940, when the development of designs and wind-tunnel models was reported to be taking place at Rostock-Marienehe. In May 1942 an aircraft that was believed to be the He 280, although no power units could be seen, was identified at Rostock. Then on 14 April 1943, a similar aircraft, possibly with a modified tail and now with two 'objects' clearly apparent on its wing, was seen at Schwechat airfield.

At the end of August 1943 it was reported that the first prototype He 280 built at Schwechat had crashed. Then the latest photographic coverage of the Austrian airfield, taken on 2 November 1943, showed four He 280s – the first time that more than one example had been seen together. By July 1944 Allied Intelligence had learnt that the development of the He 280 appeared to be considerably less advanced that that of the Messerschmitt Me 163 rocket-powered interceptor and the Messerschmitt Me 262 jet fighter. Finally, between February and 13 June 1944, some five or six He 280s were visible, with certain examples in a non-operational condition through a lack of engines or otherwise. There was no available evidence that the Heinkel fighter had been seen on airfields other than at Schwechat.

Although failing to reach production status after losing out to the more aerodynamically advanced Me 262, the He 280 did set several firsts. It was the world's first jet aircraft to have two engines, the world's first jet fighter designed as such from scratch, the first of this category to fly, the first jet aircraft to operate fitted with a tricycle undercarriage and the first jet aircraft fitted with an ejection seat, which of course was used on one occasion. Far worse designs than the He 280 have entered service in many countries. With the demise of the programme, Heinkel moved on to further jet aircraft designs such as the He 343 bomber (which never flew) and the He 162 fighter, first flown in December 1944.

He 280 (PROTOTYPE V5)	
Type	single-seat jet-powered interceptor fighter and fighter-bomber
Powerplant	2 x 1,654lb thrust HeS 8A turbojet engines (V6 2 x 1,764lb BMW 003 turbojets and V8 2 x 1,852lb thrust Junkers Jumo 004/1 Orkan turbojets)
Span	40ft 0.33in
Length	34ft 1.5in
Gross Wing Area	231.2sq.ft
Loaded Weight	9,502lb (V6 9,700lb and V8 11,475lb)
Maximum Speed	541mph at sea level, 559mph at 19,685ft for 30-second burst, 510mph at 19,685ft maintained (test pilot Fritz Schäfer reported 577mph at 19,685ft)
Initial Rate of Climb	3,757ft/min
Theoretical Service Ceiling	37,730ft
Armament	3 x MG 151/20 20mm cannon in nose (production aircraft to carry 1 x 1,102lb or 2 x 551lb bombs)

CHAPTER FIVE

GLOSTER E.28/39

E.28/39 prototype W4041/G was photographed at Brockworth just before it began taxi trials in April 1941. It is understood that after painting, the aircraft carried the serial W4041 without the '/G' on its side for its first 17 flights. The strip visible along the rear fuselage was thermal paint that had been applied to record any effects on the fuselage skin caused by the heat generated by the engine. (Author's Collection)

Once Frank Whittle had confirmed and refined the essential features of a turbojet as a potential aero engine, it was time to test it in the air. An existing aircraft type could be adapted as a test bed, but in the end the chosen route, as with both Germany and Italy, was to build an all-new aeroplane from scratch. The project was awarded to Gloster and design work was well underway before the start of World War II in September 1939. The company's design team, led by George Carter, worked on the layout with Air Commodore Whittle in what proved to be a very close and amicable partnership.

DESIGN SELECTION

Frank Whittle first visited the Gloster factory with colleague Mac Reynolds on 29 April 1939, and there they met Carter and also test pilots Maurice Summers, Michael Daunt and Jerry Sayer. Several different designs for a research aircraft were considered initially, including one that had space for gun armament in its nose and a 'tadpole-like' rear fuselage to support the tail in order to keep the latter clear of the jet stream. This mid-wing monoplane design had its jet unit positioned underneath the pilot, but it did not proceed.

The favoured layout was to be a low-wing monoplane of very clean design rather reminiscent of Gloster's earlier F.5/34 piston-engined fighter prototype, powered by a Bristol Mercury engine, that had first flown in 1937. The E.28/39 had a slightly tubby appearance and an unusually short undercarriage, since there was no longer any worry of damaging the tips of a propeller. Despite the considerable

Gloster's chief designer George Carter, seen here in 1953. He led the design team that worked on the layout of the E.28/39 with Air Commodore Whittle in what proved to be a very close and amicable collaboration. Carter headed the company's design department from 1936 to 1948, after which he served as Gloster's technical director until July 1954. (Author's Collection)

importance of this new form of jet powerplant, the designers of this diminutive aircraft wisely chose a structure that for the period was entirely conventional.

The fuselage was built in light alloy, with stressed skinning. The engine was mounted aft of the main spar, its jet pipe leading out through the monocoque rear fuselage to the extreme tail, where it ended in a propelling nozzle. Air for the engine entered through a nose intake before passing on either side of the pilot through two ducts. In order to keep the rear fuselage cool, a small amount of the air was allowed to escape past the jet pipe through a small annular hole. The nosewheel retracted rearwards into a well which filled the space below and in front of the cockpit, and an 81-gallon fuel tank was positioned immediately behind the pilot. Much of the aircraft's other fittings and some technical equipment for recording data were all accommodated within the remaining space between the intake ducts.

Due to a lack of room, the simple cockpit did not have the standard blind-flying panel installed. There were, however, some additional experimental instruments (in fact as many engine gauges as flying instruments). The pilot's view out, except to the rear, was considered to be extremely good.

The wings used stressed-skin construction around a single spar with multiple ribs, which formed a D-shaped torsion-box in conjunction with the leading edge. The Dowty main undercarriage retracted inwards to the rear of the main spar and, due to the very thin wing, both the undercarriage doors and also the wing uppersurface had to be bulged slightly to accommodate the wheels. A false rear spar supported the flaps and the ailerons, both of which had a spring-balance tab and an outboard mass-balance, and the control surfaces were originally fabric covered and operated by chain and cable controls. The tailplane had mass balanced elevators and was fitted on a fairing

A Whittle W1 jet engine stands outside the factory at Lutterworth in 1941. It was the W1 that powered W4041/G during its maiden flight in May 1941. (Science & Society Picture Library/ Getty Images)

over the end of the fuselage to the rear of the fin. A small rubber skid was placed underneath the tail to prevent any damage to the jet pipe during landing.

Other interesting features on the E.28/39 were:

a). A nosewheel steerable by the rudder. At the time this was an unusual feature, but following a forced landing it would prove very handy for manoeuvring.

b). The original W1 engine was started by an Austin Seven automobile engine connected by a flexible drive. Later on, once the early flights had been completed, the E.28/39's engine had electric starting from a ground booster battery.

c). The cockpit had a sliding canopy, but there was no pressurisation or heating of any kind. In addition, the pilot did not have a radio.

The all-up weight of the first E.28/39 with its W1 unit was 3,690lb.

The E.28/39 was designed primarily to serve as a test bed for the Power Jets/Whittle W1 engine (which, incidentally, in official documents was often called the 'Gyrone' engine) and to investigate the possibility of the jet as a means of propulsion in general. The aircraft was known as the E.28/39 from its covering specification and never received an official name. There were, however, some nicknames and, for security, several 'cover' names. It was often called the 'Pioneer' or the 'Whittle', and early official documents referred to the code name 'Weaver'. Then, before the second set of flight tests had begun, 'Tourist 1' and '2' made their appearance in reference to both prototypes performing planned flight trials out of Edgehill, in Warwickshire. 'Millet' was yet another code name used for a period, while at the start of the jet age in Britain the powerplants themselves were also often called 'Squirt engines'.

Gloster chief test pilot 'Jerry' Sayer was at the controls of E.28/39 W4041/G on its maiden flight, and for the subsequent 26 flights the aircraft made. Sadly, he was killed in an accident whilst flying a Hawker Typhoon on 22 October 1942, and never saw how successful the two E.28/39s became. (Jet Age Museum)

Specification E.28/39, dated 21 January 1940, requested a maximum speed and rate of climb at sea level of at least 380mph and 4,000ft/min, respectively. At this stage there was provision for an armament of four 0.303in Browning machine guns, although these were soon deleted. Contract SB.3229/39/C.23a was placed on 3 February 1940, and covered the building of two prototype aircraft with the serial numbers W4041/G and W4046/G. The 'G' represented Guard, signifying that the aircraft should be under guard day and night when not in use. The first E.28/39 had just the serial alone, W4041, painted on its fuselage for the first few flights.

A Mock-Up Conference was held on 22 April, and also in 1940 RAE Farnborough's Aerodynamics Department conducted low-speed wind tunnel tests on the design. The early construction work on the two prototypes was undertaken in the Experimental Shop at Hucclecote, on the edge of Gloster's Brockworth airfield in Gloucestershire. However, as the threat of German bombing raids cast their shadow over all of Britain's aircraft industry facilities, the first prototype was moved and its final assembly conducted in the Regent Motors' garage works in the centre of Cheltenham, also in Gloucestershire (the Gloster design office was also moved to Bishop's Cleeve, to the north of Cheltenham).

The first taxiing trials with W4041/G were made at Hucclecote on 7 April 1941, Whittle's own report noting how 'In the evening taxiing began in failing light. The [grass] aerodrome was so soggy that two people could not move it, though in the hanger it could be moved single-handed'. At this stage the aircraft had a Power Jets W1X engine installed, which was a 'non-flight' powerplant that had been assembled to test certain components through ground running on the bench at the test facility at Lutterworth. Consequently, it included some parts that were not intended to withstand the loads experienced during flight.

The taxi trials continued on the 8th, Whittle's own account noting that 'Sayer made three runs, leaving the ground on the first and third. On the first he got his tail down too violently and struck the blister

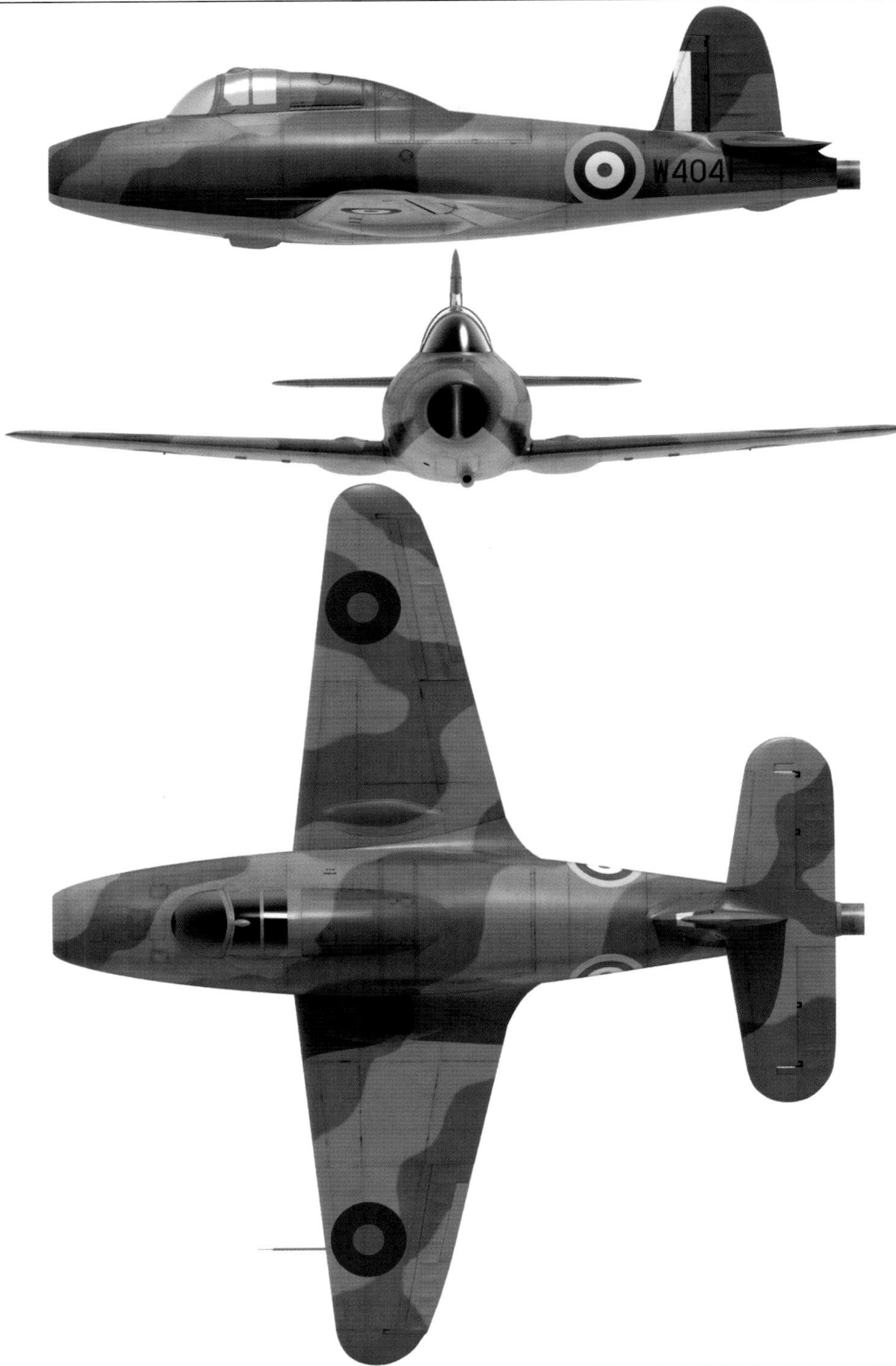

X PLANES

GLOSTER E.28/39

This three-view artwork shows the Gloster E.28/39 as it appeared on the day of its first flight. At this stage W4041 did not carry the security '/G' associated with the serials on all of Britain's early jet aircraft. The evidence confirming this is test pilot Michael Daunt's unofficial 16mm film footage of the E.28/39's first flight at Cranwell.

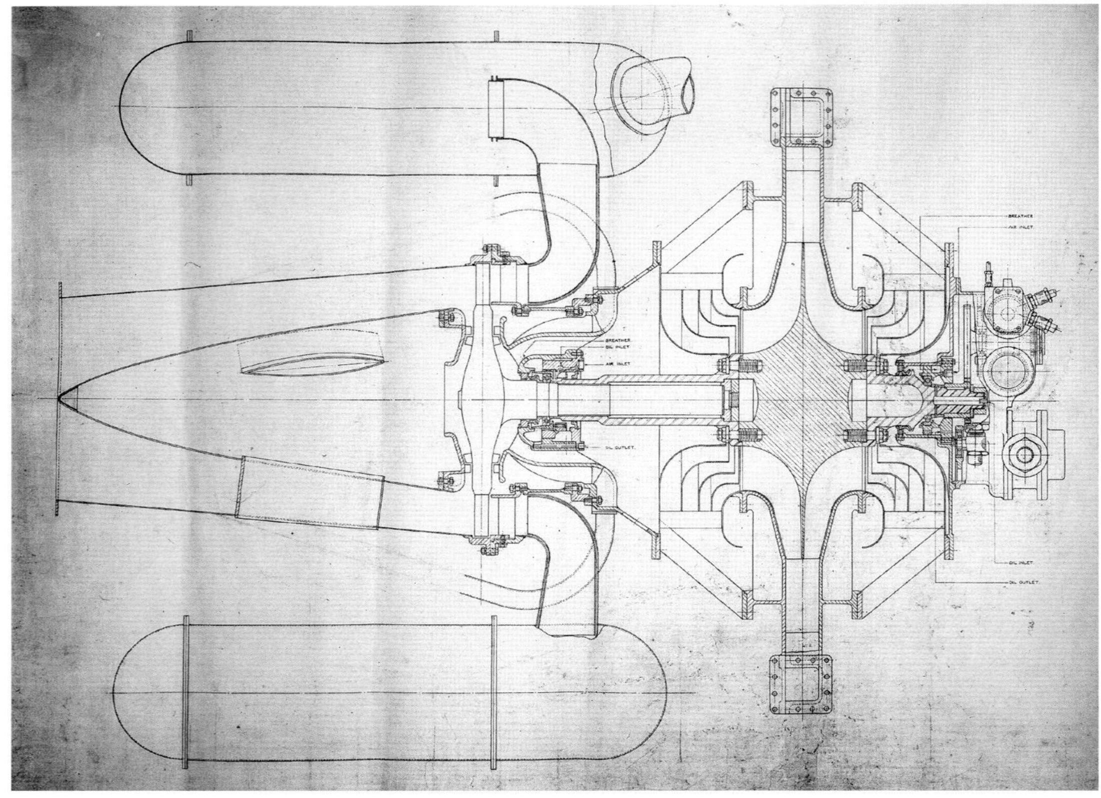

Original Power Jets drawing showing a Whittle W1A engine, an improved version of the W1 which was test flown in the E.28/39 during 1942. (Science & Society Picture Library/ Getty Images)

on the grass, which caused the machine to be thrown on its nosewheel again, thereby delaying his take-off quite considerably. On the third run he made a clean take-off and flew for about 200 yards'. Afterwards Sayer reported that he had had to force the aircraft back onto the ground, and there would have been no difficulty in making a continued flight. Now, at the time of writing 78 years later, it could be argued that these were Britain's first jet flights.

'Sayer' was Flt Lt Phillip Edward Gerald Sayer, who, since 1935, had been Gloster's chief test pilot. Many authors call him 'Gerry' after his third name, but in his 1946 book *Jet Flight,* Gloster test pilot John Grierson called him 'Jerry', and so did Gloster test pilot Michael Daunt in his logbook in April 1939. Sayer had taken the firm's F.9/37 twin-engined piston fighter aloft on its maiden flight, and he would make the first proper E.28/39 flight a few weeks later. It is understood that Frank Whittle never flew the E.28/39, but on 8 April 1941 he was one of the pilots who taxied W4041/G at Hucclecote.

AIRBORNE

On 15 May 1941 the Gloster E.28/39 became the first British jet-powered aircraft to take to the air, this landmark sortie being made from Cranwell, in Lincolnshire. Sayer was the pilot, and a flight-approved 860lb-thrust W1 engine had been installed in the aircraft before it was transported by road from Hucclecote. Cranwell had been selected as the best venue for such a flight so as to reduce the security risk associated with the early testing of this new and untried type of powerplant. The A&AEE at Boscombe Down had, for example, been turned down because of its undulating runway surface. For security purposes, the aeroplane was housed in a hangar at the extreme western end of the Cranwell airfield, and any admissions to this facility could only be made with a special pass.

The E.28/39 was in fact ready for flight on 14 May but the weather proved unsuitable, and so on that day it was used for more taxi trials. Conditions did not improve until the evening of the 15th, when Sayer took off towards the west across Cranwell's south airfield surface – the longest distance available to the pilot. In the air, he described the engine as 'quite smooth', with the noise in the cockpit resembling 'a high-pitched turbine whine'. In addition, the aircraft 'behaved normally' in gentle turns. Sayer landed again after 17 minutes.

Between 16 and 18 May speed trials were completed at 5,000ft, the performance figures coming quite close to those predicted for the engine. On the evening of 21 May the aircraft was shown to the Under-Secretary of State for Air (Harold Balfour), the Assistant Chief of Air Staff (Air Marshal Sir Richard Peck) and other officials in a very successful flying display that included a run downwind at about 1,000ft at an indicated air speed of 350mph, which was then followed by a sharp climb and a climbing turn. Originally, the elevator was found to be too light, but it was improved by fitting trimmer cord to the trailing edge, both above and below. The rudder was described as 'fairly light', with the aileron being 'reasonably light and responsive at small angles'.

With the exception of this first display, the early flying (all manufacturer's trials, and all flown by Jerry Sayer) covered general handling tests, checks for stalling speeds and then an exploration of the effects of altitude on the engine controls. The aircraft was flown by degrees at increasing altitudes up to 25,000ft, and an indicated airspeed of 300mph was reached. However, there was a stressing limit of 2g because of the low strength of the W1 engine's compressor casing. The longest flights lasted just 56 minutes because, as previously noted, the aircraft fuel tanks could only hold 81 Imperial gallons. Nevertheless, by 28 May, 17 flights had been recorded totalling 10 hours and 28 minutes in the air. When this initial set of trials had been completed the W1 engine was removed by Power Jets for inspection, and on 31 May W4041/G was returned by road to Cheltenham.

The aircraft was now grounded for nine months so that modifications could be made to the airframe and the engine replaced. Again, for reasons of secrecy, the next set of flight trials was undertaken away from Brockworth, the aircraft being transported to the airfield at Edgehill, situated roughly halfway between Gloster and Power Jets at Lutterworth. The next engine to be installed (in January 1942) was a Power Jets W1A – an improved W1. Flight trials began at Edgehill on 16 February 1942, with Sayer again the pilot. W4041/G had also now been fitted with 'GW2' high-speed, thin-section wings, but the only noticeable difference these appeared to make to the aircraft's performance over the original 'GW1' set (a NACA 230-Series section) was a 4mph increase in the stalling speed.

Some problems were experienced at Edgehill with frozen lubricating oil, which caused the engine rotor bearing to fail. Then on 27 September the E.28/39 had to make an emergency landing because of a loss in oil pressure, and in the process it suffered some minor damage. This was to be Sayer's last flight in the jet because he was tragically killed on 22 October 1942 when the Hawker Typhoon he was flying was lost in a crash.

Following Sayer's death, his assistant at Gloster, Michael Daunt, took over the E.28/39 flying programme, his first sortie on 6 November making him the second British jet pilot. Daunt's last flight before the aircraft was handed over to Service test pilots at RAE Farnborough came on 29 December 1942, during which he noted that other aircraft in the air at the same time were producing vapour trails at 25,000ft, when the E.28/39 was leaving no trail at all. The first Air Ministry test pilot to experience flying W4041/G was Wg Cdr Harold J. 'Willie' Wilson (who in 1945 established for Britain a new world air speed record of 606mph in a Gloster Meteor). An improved oil system had also been fitted for the Farnborough trials, and this proved to be satisfactory.

Gloster test pilot Michael Daunt, seen here in 1944, would become Britain's second jet pilot. He made 14 flights in W4041/G and four in W4046/G, and in March 1943 he also performed the maiden flight of the prototype Gloster Meteor, Britain first jet fighter. (Author's Collection)

The maximum speed recorded with the W1A, which delivered 1,160lb of static thrust, was 365mph. It was found that the speed did not vary greatly with height, and it was also discovered that the maximum level speed of a jet-propelled aircraft was adversely affected by changes in the air temperature – a ten-degree rise in temperature, for example, might reduce the speed by 20mph. The problem of surging, when the engine in essence appeared to 'backfire', was also encountered for the first time at high altitudes. In addition, Wilson, following several flights made during January and February 1943, criticized the E.28/39's handling qualities – in particular, there was an inadequate level of elevator control and the potential for rudder overbalance. Wilson completed this set of W4041/G trials flights on 15 February 1943, after which the aircraft was again returned to Gloster for further modifications.

W4046/G

Gloster test pilot John Grierson took the second E.28/39, W4046/G, aloft on its maiden flight, from Edgehill, on 1 March 1943. This aircraft was fitted with a 1,350lb thrust Power Jets/Rover W2B for

This very rare photograph of the second E.28/39, W4046/G, was taken in June 1943 and almost certainly at Farnborough. (Graham Pitchfork)

an all up weight of 3,950lb. This was also the first occasion that a Rover-built engine had flown, although, as explained in Chapter One, this company's involvement in jet engines was soon taken-over by Rolls-Royce. After the first flight, Grierson reported how easy the take-off was and how the subsequent climb to 1,000ft compared favourably with a propeller-driven aircraft. However, having reached a speed of 240mph he found that, on throttling back to try a stall, it seemed to take ages to lose speed compared to a propeller-driven aircraft. Landing was straightforward, but Grierson was conscious how, being so close to the ground, 'it felt as if one were travelling at a tremendous speed at the moment of touchdown'.

Summing up, he later wrote (in 1971) 'the main impressions of my first jet-propelled flight were, firstly, the simplicity of operation. The throttle was the only engine control; there were no mixture or propeller levers, supercharger or cooling-gill controls, and the fuel system had simply one low-pressure valve between the tank and the engine pump, and one high-pressure valve between the pump and the engine. There was no electric booster pump. Secondly, the absence of vibration or the sensation of effort being transmitted to the pilot's seat was outstanding – the complete reverse of the "vibro-massage" produced by the Napier Sabre's 2,400hp in the cockpit of the Typhoon'.

On 17 April 1943, Grierson made Britain's first cross-country jet flight when he took W4046/G from Edgehill to Hatfield. Two days later, piloted by Michael Daunt, the aircraft was displayed in front of Prime Minister Winston Churchill and other members of the Air Staff. Daunt made the ferry flight back to Edgehill on 20 April at 6,000ft, with the aircraft cruising at an indicated airspeed of 260mph. On this trip it was accompanied by a two-Spitfire escort, the pilots of which

Early Meteor F I EE212/G was the subject of a series of official photographs taken most likely for recognition purposes. This particular angle has been chosen to illustrate just how wide the wing nacelles had to be to accommodate the early centrifugal jet engines. (Author's Collection)

declared later 'that this seemed to be a rather fast cruising speed for their aircraft'. Grierson shared the second aircraft's development flying with fellow company test pilot John Crosby Warren (who on 27 April 1944 was killed when the prototype Gloster Meteor F.9/40 he was flying had a trim tab detach, causing the aircraft to roll inverted into the ground). In April 1943 the development of the W2B engine was passed to Rolls-Royce Ltd.

On 3 May 1943, W4046/G went to RAE Farnborough to undergo a programme of intensive flying. Here again it was observed that the engine thrust when landing was quite high, which in turn made the approach more difficult than it would have been in a propeller aircraft. As noted previously by Grierson, this also had the effect of making the speed of a 'glide' from altitude very high, although this characteristic was due in part to having no propeller to act as an airbrake (a clean propeller-less aeroplane picked up speed very rapidly). A speed of 435mph was subsequently achieved in W4046/G and a height of 35,000ft was reached in 27 minutes after take-off. The stalling speed (flaps down) proved to be 65mph, and the aircraft could be flown flaps up at 85mph.

On 4 June 1943, the RAE's Grp Capt Allen H. Wheeler took W4046/G on a flight specifically to see how it behaved under aerobatic conditions. He described this flight as follows:

'The aircraft was taken off and climbed to 7,000ft to find the exact cloud base and then tested in rapid banks either way, and the general handling up to 3g was tried out in turns. The aircraft was then dived to 4,000ft and pulled up, with the nose about ten degrees above the horizon, and a roll was done to the right at 250mph. The aircraft went over very smoothly indeed, and is probably one of the nicest aircraft to roll I have yet had. Immediately afterwards I dived the aircraft to about 280mph and executed two more rolls to the right in immediate succession about 20 degrees above the horizontal. The aircraft maintained approximately 250mph at the end of this, and again was very easy [to fly].

'I then dived the aircraft to 2,000ft at 350mph and pulled it up at between 2.5 and 3g to do a loop. I eased off the "g" to about 1¾ at 6,500ft as I was in the inverted position, doing approximately 140mph. As the aircraft passed over the vertical it shook violently aerodynamically, as all modern aircraft do if they go round too fast at the top, but it immediately straightened out again and completed the loop. I eased back the throttle slightly at the top and the loop was completed without any difficulty.

'This aircraft would be perfectly easy to loop like any other, except in this particular case I was unwilling to go into the cloud [at 7,000ft] at the top off the loop and pulled it back rather faster than I would normally like to do. This would account for the aerodynamic shake, which in fact was no worse than any Spitfire would do under the same conditions. The actual loop can be executed without any rudder at all, and is perfectly straightforward as far as all operations are concerned. The engine appeared to be unaffected by the aerodynamic shake and ran perfectly smoothly afterwards.

'I then did two more rolls 30 degrees above the horizontal at 260mph, followed by one to the right, and then turned and did another roll under the same conditions to the left. These further rolls confirmed my previous view that it is a very pleasant aircraft to roll. Throughout all these manoeuvres I noticed no fluctuations in engine power or vibration from it at all, nor did the jet temperature vary, except in the way it normally does with change of burner pressure. The highest speed recorded was 360mph, and no rudder instability was noticed, although this speed was attained at the bottom of the dive in rather bumpy air when a small amount of instability would not be noticeable.

'In conclusion, I should like to say that I consider this to be a very excellent aircraft and engine indeed to do aerobatics, and I do not anticipate any trouble at all in executing any manoeuvres.'

Wheeler's positive remarks notwithstanding, he and his fellow test pilots at the RAE had found that W4046/G's rudder control was much too sensitive, and it was considered that the fin area should be increased.

A few days after this flight a more powerful 1,520lb thrust W2B engine was installed in W4046/G, Wilson making the first flight with this unit on 20 June. A speed of 466mph was later achieved at 10,000ft and the time to climb to 35,000ft was cut to 26 minutes. During further aerobatic flights, particularly by Sqn Ldr Charles G. B. McClure, it was found that the fuel tank was susceptible to negative 'g', which prevented fuel from reaching the engine – however, it did prove possible to restart the engine in the air. W4046/G would also 'zigzag' slightly at high speeds.

On 30 July 1943, Sqn Ldr Douglas B. S. Davie took the second E.28/39 to an altitude of 33,000ft, at which point the aileron controls jammed and he was forced to 'bail out' – the glass of the aircraft's canopy shattered and the gyrating jet jettisoned him into a 20,000ft freefall. Davie lost his boots, helmet and oxygen mask in the process, although fortunately for him he was able to breathe by sucking on his severed oxygen tube and successfully open his parachute. He survived what was probably the longest parachute descent recorded up to this time with only a mild touch of frostbite. Sadly, Davie was another test pilot who would subsequently die in the loss of a Meteor prototype (on 4 January 1944) due to mechanical failure – an engine disintegrated on a high-speed test run at 20,000ft.

W4046/G crashed near Shalford, in Surrey, and an analysis of the wreckage indicated that the most probable cause of the jammed ailerons was quite literally that they had frozen up due to freezing water both on the surfaces and in the control circuits. It was recommended that W4041/G should be modified to prevent this happening again. Altogether, W4046/G had made 139 flights for a total of 69 hours and 36 minutes in the air.

W4041/G

On 23 May 1943, W4041/G began new flight trials with a 1,520lb Power Jets W2/500 engine in place. This flight was also the first time that the E.28 had flown out of its 'home' airfield at Brockworth (something W4046/G would never do). Between 12 and 26 June it flew from another temporary home at Barford St John, in Oxfordshire, Daunt and Grierson sharing the flying. In fact, it is quite extraordinary that these two aircraft completed so little flying from the airfield where they were built, W4041/G making just seven flights in all from Brockworth. The great majority of their flying of course took place out of Farnborough.

The aircraft's performance was now quite similar to W4046/G with the W2B installed, and a height of 41,600ft was achieved (by Grierson on 24 June 1943) without any surging, although the pilot found the extreme cold made it very difficult to write down the necessary performance figures. This was the first occasion that 40,000ft had been reached by a jet-propelled aeroplane. Daunt declared the new turbojet, fitted in what was described as a rejuvenated '4041', to be 'the smoothest unit that this pilot has had the pleasure of flying'.

It was also during this new series of flights that a vapour trail – possibly the first to be produced by any jet-powered aeroplane in Britain at least – was seen. Hitherto, it had been considered that, due to the tremendous heat and volume of the jet efflux, a jet-propelled aeroplane would not produce a vapour trail. However, during Grierson's record altitude flight on 24 June the contrary was proved when an observer at Brockworth noticed quite clearly a vapour trail as he passed overhead at around 36,000ft.

With the loss of the second aircraft, the first E.28 had to be grounded for modifications to be made to its aileron control circuit. The wing section was replaced by a new Gloster wing of special high-speed 1240 section at the same time, and there was a new-build tailplane incorporating small inset finlets to cure the 'zigzagging' experienced at

A later view of W4041/G taken at Farnborough. As built, the aircraft stood with its fuselage axis very near to horizontal and with its low-set wing very close to the ground (partly because the new jet engine did not need any clearance for a propeller). This may well have limited the amount of wing lift. However, before W4041/G first flew from Cranwell, and for an entirely different reason, Crabtree's Garage in Cheltenham (a Gloster dispersal site) extended the nosewheel leg. As a result, there were never any problems with initial rotation during take-offs. It also meant that when on the ground, the E.28 had a distinct nose-up stance. In this view the first E.28 has had additional tailplane finlets fitted. (Author's Collection)

high speeds. A jettisonable cockpit hood was also introduced, although these changes took some time to complete because Gloster was now very busy with the production of its Meteor jet fighter.

The stimulant that brought a change of role for W4041/G, with the introduction of these various modifications, went back to 1943 when RAE Farnborough stepped up the pace of its flight trials and research into high Mach numbers. The only way the necessary data could be garnered was to dive a test aeroplane steeply until the required high speed had been reached, and then the effects of compressibility on its drag, stability and control could be measured. Several aircraft were employed in these experimental trials, which in 1944 included the famous occasion when Sqn Ldr Anthony F. 'Tony' Martindale took his Spitfire IX to a true Mach number of 0.9.

W4041/G's role at RAE Farnborough came under the control of the RAE's T Flight ('T' for Turbine). However, by September, the aircraft had returned to the strength of A ('A' for Aerodynamics) Flight. In truth, the E.28/39 was no longer a pioneer of general turbojet flying. From now on it was to be engaged in the research into the problems of 'compressibility' (the term used to cover the point where shock waves first appeared and affected an aircraft's controllability) during high-speed flight, which would also involve its pilots diving W4041/G to high Mach numbers.

Flying resumed at Farnborough on 9 March 1944 with a more powerful W2/500 engine installed, although it was soon found that this unit was prone to surging at 25,000ft and above so it had to be modified. On 25 April the aeroplane was climbed to 42,710ft by Sqn Ldr B. H. Moloney, which may have been the highest altitude ever reached by W4041/G. On this occasion the aircraft could have gone

Attached to an external auxiliary power cart (out of shot to the right) probably at Farnborough, E.28/39 W4041/G was photographed between flights in 1944. (Science & Society Picture Library/Getty Images)

higher – the pilot noted that the jet still had an estimated rate of climb of 1,000ft/min, so this was not its ceiling. However, as the cockpit was not pressurised, this was, therefore, the limit for the pilot. In fact the RAE's pilots took W4041/G to heights greater than 40,000ft on several different flights. Also on this 25 April flight Moloney reported that, as he passed through 33,000ft on his return to base, he recorded a Mach number of 0.83.

A maximum level speed of 480mph was now reached, with the all-up-weight (due to the addition of more equipment) rising to around 4,200lb. The potential of the W2/500 installation, with its extra power, should have enabled the E.28/39 to demonstrate, at last, a much more impressive performance perhaps similar to that first envisaged in 1939. Unfortunately, this engine's weakness with surging at higher altitudes prevented the pilots from taking advantage of the more favourable conditions found in the upper atmosphere.

The next stage of this 'compressibility' programme saw the first example of yet another different engine, the W2/700, being fitted into W4041/G in July 1944. This type, with a thrust rating of more than 2,000lb, was being prepared for the forthcoming Miles M.52 supersonic research aircraft (which was later cancelled before it had flown). The W2/700 had been developed directly from the W2/500, and it was the last of Frank Whittle's own engine designs.

To prepare the E.28/39 for its next round of flights, and to allow the airframe to cope with the considerable extra power of the W2/700 and to try and ensure that high-speed dives and recoveries would not damage the aircraft in any way, the wing structure had to be strengthened – particularly around the bulges over the main undercarriage wells. In addition, Gloster now replaced the fabric covering on the ailerons and elevators with metal-skinned versions. The surviving E.28/39 was also modified by the addition of a wing-mounted pitot comb and other special instruments.

To gather data, on a typical test flight the aircraft would be flown to a high level speed and taken close to its ceiling. Then the pilot would nose over into a power dive, usually at around 45 degrees but on occasion as much as 60 degrees. This had then to be held for a sufficient period to allow the speed to stabilise, during which time height would be lost very swiftly indeed, before the pilot could commence a pull-out at 2–3g. Like most other aircraft taking part in these trials, as the E.28/39 increased its speed in the dive the jet experienced an increasing nose-down change of trim.

The first flight with a W2/700 in place was made on 25 August from Farnborough. During this period some very high Mach numbers were recorded by Moloney and the famous RAE test pilot Cdr Eric Brown. One characteristic to surface that affected the E.28 was pitch oscillation of between 0g and 4g. Indeed, in some dives a violent pitching motion was experienced, and to suppress them the elevator trailing edges later had to be 'blunted' by cementing on strips of 3/16in diameter cord. The W2/700 proved to be less affected by surging, but once again it was not possible to take full advantage of the potential offered by an engine that provided far more power than the original W1. This time the rear fuselage structure would not permit an optimum diameter of jet pipe to be fitted to match the power output.

Nevertheless, the Mach number was increased progressively until a figure of Mach 0.816 was recorded on 6 December 1944. Some sources state that this was the E.28/39's highest ever speed, although other published accounts report that on 27 August 1944 Brown reached Mach 0.93 in a dive. What is known is that at these speeds W4041/G 'became extremely difficult to handle, due to buffeting and general oscillations about all three flight axes'. In fact Frank Whittle became quite worried that the only surviving E.28/39 might be lost, and in June 1944 he presented a strong protest against W4041/G being used in these trials to Sir Ralph Sorley, Deputy Director of Research and Development at the Ministry of Aircraft Production. Sorley did not agree, so in his next memo Whittle pushed a case for constructing 'two or three more E.28s'. It should be added that with these more powerful engines the E.28/39 apparently recorded ground level rates of climb of more than 5,000ft/min – at the time, a very high figure.

This final trials programme with the E.28/39 enabled a great deal of valuable data to be acquired regarding flight at high speeds and altitudes. Into 1945, sorties during January and February were limited by problems with the engine. In all, three different W2/700 units were installed at different times, although only a modest amount of flying was done with all of them. W4041/G's very last sortie was made on 20 February 1945, and in a total of at least 100 flights this aircraft had recorded 51½ hours of flying time.

As an aside, one new problem connected with the operation of the jet engine in these early days was the supply of fuel – kerosene (paraffin). Only those test airfields in Britain from which jet aircraft were currently operating would stock it, so an engine failure followed by a forced

During its flying career W4041/G had small finlets fitted to the outer part of each horizontal tailplane. Later, after retirement and in readiness for public ground displays staged in 1945, the E.28/39 was refurbished and these finlets were removed once again. The machine was now as close to its original form as was possible (apart from the inscription on the nose). This photograph of the first British jet aeroplane to fly was probably taken at Brockworth in 1945 just after the restoration had been completed. (Author's Collection)

landing somewhere else meant that the jet-powered aeroplane could not be refuelled to make a return flight. The solution was to fly in an Avro Lancaster bomber fitted with a fuselage-mounted tank full of jet fuel (plus the necessary maintenance crew to service the aircraft and the security guards to screen it).

UNVEILED

On 6 January 1944, a joint statement was issued by the US War Department in Washington, D.C. on behalf of both the RAF and the USAAF, which revealed for the first time the existence of 'the jet-propelled Gloster monoplane' and, of course, the development of the jet engine itself. As *Flight* magazine observed in its next issue, this had been the most well-guarded secret 'because no hint had ever been given in any English journal that this country was even experimenting with jet propulsion'. And yet it had also been the best-known secret because thousands of people along the length and breadth of Britain had seen and heard the aeroplane, while many more were directly involved with it.

After the E.28/39 had begun its flight test programme in Lincolnshire, the magazine started to receive dozens of letters from readers living in that part of the country, all asking 'what this new weird aircraft was which flew about at terrific speeds and had no airscrew'. Until now, the secrecy regulations had prevented the press from saying anything, but at last the secret was out.

Many years later it is quite difficult to imagine what members of the general public, conditioned to seeing only piston-engined aircraft or gliders and hearing engines that drove propellers, thought about these

E.28/39 W4041/G has been on public display in the Science Museum in South Kensington, west London, since April 1946. (Jet Age Museum)

jet-powered machines that made a strange whining noise and did not have any propellers at all! Seeing them must have been puzzling to say the least, and probably quite an extraordinary experience. With the security issues now eased, on 11 December 1944, W4041/G was specially flown by test pilot Cdr Eric Brown on a photographic sortie to ensure that an image record of the E.28/39 was made before the aircraft was retired.

In its original form, the E.28/39 could never have been turned into a fighter aircraft. Nevertheless, some consideration was given to producing a fighter version, and prototypes of such a development were ordered against Specification E.5/42. These were never built in their original form, but after further development and modification, a single-engined jet fighter prototype was flown on 9 March 1948 as the Gloster Ace. However, this aircraft exhibited a poor performance and, by the time it appeared, was already so out of date that it did not progress further. This was not really a problem for Gloster because well before the end of the war the firm had been working on its twin-engined Meteor jet fighter, the prototype of which first flew on 5 March 1943. This aircraft became a huge success, remaining in production in numerous versions for both the home market and abroad until well into the 1950s.

In March 1945 the surviving E.28/39 was re-allotted to Gloster at Bentham, in Gloucestershire, for what was termed 'reconditioning', instructions having been given to remove all of the airframe's experimental equipment. Then between 21 June and 16 September it was included in a display of historic aircraft staged by the Ministry of Aircraft Production in London's Oxford Street in what was called the 'Britain's Aircraft' exhibition. Afterwards, Gloster reportedly used

W4041/G in several more ground exhibitions – it is certainly known to have been at the company's Bentham works in March 1946. Finally, on 24 April 1946, the E.28/39 was moved from Brockworth to its last resting place, the Science Museum in South Kensington, west London, to go on permanent display, and it has remained there to this day.

It was fitting that this remarkable little aeroplane should be housed in a place of honour as a tribute to all those designers, engineers and pilots whose initiative and energy had ensured its success.

Although the last of the aircraft described in this book to make its maiden flight, the E.28/39 was undoubtedly the most successful. It quickly fulfilled its initial purpose of demonstrating the practicality of the very new form of jet propulsion. Next, it became an important tool in the development process for the first generation of gas turbine engines. Finally, because it possessed splendid aerodynamic qualities, it was able to take on the new and altogether different task of a research aircraft for flight at high subsonic speeds. The two examples built were also used to give as many pilots as possible their first experience of jet flight. After completing such an exceptional career, the E.28/39 can be looked upon as one of Britain's most successful research aeroplanes.

E.28/39	
Type	experimental single-seat jet-powered aircraft
Powerplant	1 x Power Jets turbojet engine (see text for various installations)
Span	29ft 0in
Length	25ft 3.75in
Gross Wing Area	146.5sq.ft
All Up Weight	varied with different engines (see text)
Maximum Speed	480mph with W2/500, possibly Mach 0.93 in a dive
Ceiling	uncertain (varied with different engines – see text)
Armament	none fitted

BIBLIOGRAPHY

Alegi, Gregori, *Campini, Caproni and the C.C.2*, The Aviation Historian Issue 6 (January 2014)

Brinkworth, B. J., *On the Aerodynamics of the Gloster E.28/39 – a Historical Perspective*, Aeronautical Journal of the Royal Aeronautical Society (June 2008)

Buttler, Tony, *British Experimental Combat Aircraft of World War II*, Hikoki (2012)

Desoutter, D. M., *Aircraft and Missiles*, Faber & Faber (1959)

Green, William, *Warplanes of the Third Reich*, Macdonald and Jane's (1979)

Green, William, *Harbinger of an Era – The Heinkel He 280*, Air International (November 1989)

Grierson, John, *Britain's First Jet Aeroplane*, Flight (13 May 1971)

Johnson, Brian, *The First of the Jets*, Aeroplane Monthly (July 1992)

Jones, Glyn, *The Jet Pioneers: The Birth of Jet-Powered Flight*, Methuen (1989)

Kay, Antony L., *Turbojet History and Development 1930–1960 Volume 1: Great Britain and Germany*, Crowood (2007)

Kay, Anthony L., *Turbojet History and Development 1930–1960 Volume 2: USSR, USA, Japan, France, Canada, Sweden, Switzerland, Italy and Hungary*, Crowood (2007)

Kershaw, Tim, *Jet Pioneers: Gloster and the Birth of the Jet Age*, Sutton (2004)

Matthews, Henry, *Gloster-Whittle E.28/39 Pioneer: A Flying Chronology*, X-Planes Profile 3 (2001)

Schäfer, Fritz, *I Flew the World's First Jet Fighter*, RAF Flying Review (October 1958)

Schick, Walter and Meyer, Ingolf, *Luftwaffe Secret Projects: Fighters 1939–1945*, Midland Publishing (1997)

Smith, G. Geoffrey, *Gas Turbines and Jet Propulsion*, 6th edition, Iliffe (1955)

Smith, J. R. Smith and Kay, Antony L., *German Aircraft of the Second World War*, Putnam (1990)

Thompson, Jonathan W., *Italian Civil and Military Aircraft, 1930–1945*, Aero Publishers (1963)

Whittle, Air Commodore Frank, *The Early History of the Whittle Jet Propulsion Gas Turbine, 1st James Clayton Lecture*, Institute of Mechanical Engineers (October 1945)

Other sources consulted included the *Flightglobal Archive* website (*Flight* magazine archive) and the British National Archives at Kew, the files examined at the latter venue including Whittle's own papers in Record Group AIR 62.

INDEX

Note: page numbers in bold refer to illustrations and captions.